MW01029355

Aerodynamics
of the
AIRPLANE

Aerodynamics
of the
AIRPLANE

CLARK B. MILLIKAN

DOVER PUBLICATIONS
Garden City, New York

Bibliographical Note

This Dover edition, first published in 2018, is an unabridged republication of the work originally published by John Wiley & Sons, Inc., New York, in 1941.

International Standard Book Number

ISBN-13: 978-0-486-82370-6
ISBN-10: 0-486-82370-9

Manufactured in the United States of America
82370904   2022
www.doverpublications.com

EDITORS' PREFACE

In the nearly thirteen years since the Guggenheim Aeronautical Laboratory at the California Institute of Technology was established, a large number of scientific and engineering problems in various fields related to aeronautics have been investigated under its auspices. The results of this activity have appeared in the form of numerous technical papers. However, it has been felt for some time that certain of the subjects dealt with at the laboratory deserved more systematic treatment. It has also been suggested that aeronautical texts based on the staff's experience in training aeronautical engineers and scientists might be of some value. Accordingly, about two years ago an agreement was made by the undersigned editors with John Wiley and Sons, Inc., for the publication of a series of volumes, covering these two fields, to be known as the GALCIT * series. A start was then made on the preparation of material for certain of the monographs of this series, but the press of work associated with national emergency defense activities and the rapidly changing nature of the GALCIT'S academic courses made progress very slow.

In the spring of 1940 the Lockheed Aircraft Corporation requested the California Institute to cooperate in its expansion program by giving a course of aeronautical training to a large group of graduate non-aeronautical engineers. This program was undertaken during the following summer, when it appeared that the material presented, being somewhat more general than that given in the regular GALCIT postgraduate courses, might fill a rather important gap in the aeronautical training literature. The lecture notes were accordingly worked over into material for three volumes dealing with various aspects of aeronautical engineering. These three volumes, of which the present is the first, constitute the initial publications of the GALCIT series.

Acknowledgment must be made of the cooperation extended by the Lockheed Aircraft Corporation which released much of the information

---

* The rather unwieldy title "Guggenheim Aeronautical Laboratory, California Institute of Technology" has been abbreviated to the more convenient "GALCIT." This notation has been widely used in aeronautical circles and is employed throughout the present work.

originally issued for use in the training course, and whose stimulus was
responsible for the preparation of the initial lecture notes on which the
volumes are based. The Douglas Aircraft Company was also most
cooperative in furnishing valuable material for incorporation in these
books.

THEODORE VON KÁRMÁN
CLARK B. MILLIKAN
Editors

April, **1941**

AUTHOR'S PREFACE

The present treatment of aerodynamics was undertaken, as indicated in the Editors' Preface, for the purpose of presenting a brief but rather intensive summary of those portions of the subject which should be included in the background of knowledge of any well-rounded aeronautical engineer. In view of the fact that the work was not intended as a reference volume for aerodynamic specialists an unusual emphasis has been placed on fundamental fluid mechanical principles, while material of the handbook type has been to a large extent excluded. Methods of performance estimation and analysis have been treated in some detail and two working charts included, since it was felt that the non-specialist might most frequently have occasion to make estimates or calculations in this branch of aerodynamics.

The material in the volume has been presented in a form suitable for students whose knowledge of engineering or physics includes only a familiarity with the elements of mechanics, and whose mathematical experience extends only to the simplest fundamentals of differentiation and integration. However, an attempt has been made to indicate certain of the problems whose complete treatment involves more elaborate mathematical techniques, in the hope that the book might furnish a satisfactory and perhaps stimulating introduction to students who might subsequently enter more deeply into the field of aerodynamics.

The author has drawn heavily on the technical publications of the National Advisory Committee for Aeronautics, and wishes to express his appreciation to the committee for its permission to use such material, and especially for its courtesy in making available drawings and photographs used in several of the figures. Grateful acknowledgment is also made to the McGraw-Hill Book Company for its kindness in furnishing reproductions of the flow photographs which appear in the volume. Finally the author is happy to express his indebtedness to the many members of the GALCIT staff with whom he has had the privilege of discussing the subject matter of the volume and who have assisted in its preparation. Mr. E. W. Robischon in particular has contributed greatly to the completion of the manuscript.

CLARK B. MILLIKAN

CALIFORNIA INSTITUTE OF TECHNOLOGY
December, 1940

CONTENTS

ix

INTRODUCTION

Of the many types of problem dealt with by the aeronautical engineer, two combine to constitute the major portion of the field called *Aerodynamics of the Airplane*. The first is concerned with the supporting of weight in the atmosphere and the raising of this weight to altitudes above the earth's surface, as well as with the moving of the supported weight horizontally through the atmosphere. In order to be called an airplane the machine employed for these purposes must be heavier-than-air, it must, when in flight, have no material connection with the earth other than that afforded by the atmosphere itself, and its supporting surfaces must be essentially stationary with respect to the weight carried. All questions connected with the magnitude of the weight lifted, the height and distance through which it is carried, and the speed with which the motions take place are included in the subject of *Airplane Performance*. The second type of problem includes all questions involved in the maintaining of stable and controlled flight, so that the objectives considered in the subject *Airplane Performance* may be satisfactorily and safely attained. Such questions are grouped together under the title of *Airplane Stability and Control*.

Unfortunately it is not possible to treat any of the important problems which occur in either of these two branches of *Aerodynamics of the Airplane* without a rather detailed preliminary study of the fundamental relations involved. Furthermore, many of these relations appear at first sight to have very little connection with the practical concerns of the aeronautical engineer. The first chapter of the present treatment of the subject accordingly deals with general principles and experimental data, and much of its content may at first seem somewhat abstract and academic. The second chapter discusses the application of certain of these fundamental relations to specific aerodynamic questions, and reviews selected portions of the enormous body of experimental material which aerodynamic laboratories have made available in the last few years. *Performance* is treated in the third chapter and two aspects of *Stability and Control* furnish the subject matter of the last two chapters.

CHAPTER 1

FUNDAMENTAL PRINCIPLES

1–1. Aerodynamic Forces; Dimensionless Coefficients

Aerodynamics is one branch of the general subject of fluid mechanics which is to a great extent concerned with the reactions occurring between fluids and solids immersed in them.   Before proceeding to a discussion of the more specialized field it will be helpful to consider briefly the general question of the forces which act on a body when it moves through any fluid.   Here a very elementary relativity principle is

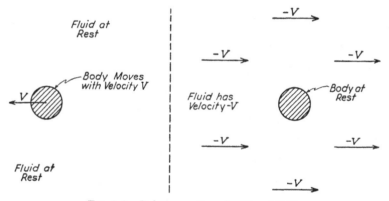

FIG. 1·1.   Relative motion of solid and fluid.

useful.   This principle states that the force exerted by the fluid on the body is exactly the same in the following two cases illustrated in Fig. 1·1:

   (a) The body moves with a velocity V through the fluid which is at rest far from the body.
   (b) The body is at rest and the fluid flows past it, the fluid far from the body moving with a velocity $-V$.

The principle may be stated somewhat more concisely in the form: *The force exerted by a fluid on a body does not depend on the absolute*

1

velocity of either fluid or body, but only on the relative velocity between them. In aeronautics we are usually concerned with the force on an airplane or a part moving through the air which is considered to be at rest far from the airplane. For purposes of visualization, however, it is often more convenient to think of the airplane as at rest with the entire atmosphere moving past it. The relativity principle states that the force on the airplane or part will be identical in the two cases.

With this in mind consider an element of solid surface, ΔS, at rest with a fluid flowing past. The element is taken so small that it may be considered as flat. Figure 1·2 represents a plane perpendicular to ΔS; the fluid is assumed to move in the plane of the paper. The force

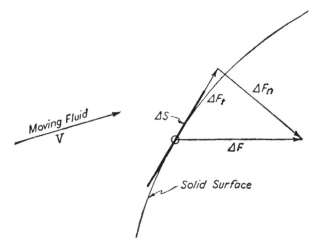

Fig. 1·2. Force components on a surface element.

which the fluid exerts on ΔS may be represented by the vector ΔF, giving both the magnitude and the direction of the force. It is found convenient to resolve this force into its two components normal and tangential to the surface element ΔS. These components are indicated by ΔF_n and ΔF_t. If the force components are divided by the area, ΔS, of the element, quantities are obtained which have the nature of force per unit area, or stress. The normal stress, $\Delta F_n/\Delta S$, is called the average pressure, p, over the element, and the tangential stress, $\Delta F_t/\Delta S$, is referred to as the shearing stress, τ. The values of p and τ at a point on the surface are obtained by determining the limits which the above ratios approach as the area, ΔS, surrounding the point goes to zero.

Later it will be seen that the pressures and shearing stresses at the surface of a body moving in a fluid have very different physical origins, so that it is convenient to separate the resultant forces on the body arising from these two types of stress. The so-called " pressure force " is obtained by adding the force vectors ΔF_n due to the action of pressure perpendicular to every element ΔS of the surface of the body. Since each ΔF_n is a vector, having both magnitude and direction, the addition must be made vectorially so that the total pressure force is actually the resultant of all the elementary forces ΔF_n. Similarly the resultant of all the elementary tangential forces ΔF_t may be obtained and is called the friction force or " skin friction " acting on the body.

The question now arises of determining and making available to the engineer data giving the resultant forces which will occur on the moving bodies with which he must work. The most direct procedure would be to set laboratories to work determining such forces under all the various conditions which could be expected to occur. The results could then be presented tabularly in some sort of handbook form. It is at once obvious that the enormous amount of material which would be required to present the results would make such a collection practically useless. What is first needed is some sort of systematization which will permit an orderly presentation in terms of the various parameters of greatest importance. Such a systematization was accomplished by the great English engineer and physicist, Osborne Reynolds, who laid the foundations for modern fluid mechanics during the latter part of the nineteenth century.

The most important of Reynolds' results in this connection can be obtained quite easily through the application of elementary dimensional analysis. In order to demonstrate this the problem is simplified as follows: One member of a family of geometrically similar solid bodies is considered. The body moves in a given direction with uniform velocity through a homogeneous fluid which extends to infinity and is at rest except for the velocities produced in it by the motion of the body. The force exerted by the fluid on the body as a result of the motion is to be found. In order to proceed all the parameters which can enter into the problem must first be listed. Gravity cannot have any effect since the fluid is homogeneous and any forces due to buoyancy will be independent of the motion, having the same value whether the body is at rest or moving. Since gravity does not enter the problem the various members of the family of geometrically similar bodies are characterized only by their size. In fact, if some characteristic linear dimension l is chosen, e.g., the diameter if the bodies are spheres, the length if they have streamline or fuselage shape, then any member of the family is completely defined by a numerical value for l. The velocity with which the body moves is denoted by V. The force F on the body will then depend on l, V, and on the properties of the fluid through which the body moves. This dependence can be indicated symbolically in the form:*

$$F = F(l, \ V, \ \text{fluid properties}) \qquad [1\cdot1]$$

The parameters which define the fluid properties must be found. The first is obviously the density which is universally written as ρ = mass of fluid per unit volume.† The second is connected with the shearing stress which occurs when adjacent layers of fluid slide past one another. The simple law governing this phenomenon was originally given by Newton and has been so often verified experimentally that it may be considered as an empirical law. It can be understood most readily in connection with the following very simple type of flow. The motion is taken as two-dimensional which means that the velocity at every point has a direction lying in a plane parallel to a fixed reference plane (as, for example, that of the paper). Furthermore the distribution of velocities is identical in all of these planes. The motion is further restricted in that, of the two velocity components in such a plane (u parallel to the x axis and v parallel to the y axis, using the coordinate system indicated in Fig. 1·3) one is required to vanish. We choose the coordinate system so that the v component vanishes and the velocity at any point is completely determined by the value of u at the point.

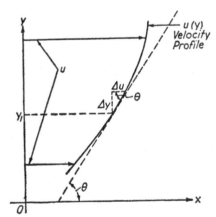

FIG. 1·3.   Velocity profile notation.

The type of diagram shown in Fig. 1·3 is very convenient as a means of visualizing the flow conditions and will be used frequently throughout the later discussions. It represents the distribution of velocity over a section x = constant, taken as $x = 0$ in Fig. 1·3. At each value of y along this line a vector is drawn parallel to the velocity, i.e., to the x axis, the length of the vector being proportional to the magnitude of the velocity at the point. The curve joining the end points of all these vectors is called the *velocity profile* and gives an

---

* Throughout the work we shall use the notation $z = z(x, y, \cdots)$ which is to be read "z is a function of $x, y, \cdots$."

† This is not to be confused with the specific weight, γ, which is usually used by hydraulic engineers. $\gamma = \rho g$, where g = acceleration of gravity.

immediate visual representation of the dependence of u on y, written symbolically $u(y)$.

Newton's friction law mentioned above gives the relation connecting the shearing stress between two adjacent layers of fluid at an arbitrary height, say y_1, and the velocity distribution $u(y)$. It states that this shearing stress is inversely proportional to the slope of the velocity profile, or expressed analytically

$$\tau = \mu \frac{du}{dy} \qquad [1\cdot2]$$

where $\mu =$ a constant of proportionality, and

$\dfrac{du}{dy} =$ the velocity gradient = the rate of change of u with respect to $y = \dfrac{1}{\tan \theta}$, where $\tan \theta =$ the slope of the velocity profile at the value of y in question (in this case y_1).

The constant μ is found to depend on the nature and state of the fluid and is called the coefficient of viscosity. It is large for sticky or viscous fluids like lubricating oil and small for " watery " fluids like water itself or air. It is actually the physical constant which determines the nature of any fluid with regard to shearing frictional phenomena.

A third physical property of the fluid which might be expected to be of importance is the compressibility. Liquids like water are, however, so nearly incompressible that this factor does not enter, and even for gases like air compressibility has a negligible effect for many important phenomena. We shall later investigate a little more closely the conditions under which compressibility can be neglected and shall for the present assume that the fluids under consideration may be treated as incompressible.

It appears that the properties of the fluid are completely given for the purposes of this study by the values of the two parameters ρ and μ. The symbolic equation 1·1 may then be written

$$F = F(V, l, \rho, \mu) \qquad [1\cdot3]$$

The relation between the force F and the four independent parameters, V, l, ρ, μ, can now be given a much more precise form through the application of the principle of dimensional homogeneity. This principle may be stated very crudely as follows: " In any equation of the form

$$A + B + C + D + \cdots = M$$

each of the terms $A, B, C, D, \cdots, M$ must have the same dimensions if the equation has any physical significance."

No proof of this principle will be attempted here, but its fundamental importance in two quite different connections must be emphasized. First its application leads to the relatively simple derivation of many fundamental physical laws like that below. Second its continued use during the analysis of any physical problem can be of immense assistance in bringing to light errors which might otherwise be very difficult to discover. This application of the principle is especially valuable in routine aerodynamic calculations.

For purposes of dimensional analysis a very convenient notational convention is introduced, namely, the surrounding of a symbol by bars to indicate that only the dimensions of the quantity are to be considered. Thus $|A|$ is to be read *the dimensions of A*. With this convention the application of the principle of dimensional homogeneity to the equation above would lead to the set of equations

$$|A| = |B| = |C| = |D| = \cdots = |M|$$

In applying the principle to equation 1·3 a more useful form for the function F is needed. A very general expression is

$$F = \sum_i C_i \rho^{a_i} \mu^{b_i} V^{c_i} l^{d_i} \qquad [1\cdot4]$$

where the exponents $a_i \cdots d_i$ can have any values and the coefficients C_i are dimensionless numbers. The symbol $\sum_i$ means that all the terms for any number of values of the index i are to be added. Writing out a few of the terms results in

$$F = \cdots C_1 \rho^{a_1} \mu^{b_1} V^{c_1} l^{d_1} + C_2 \rho^{a_2} \mu^{b_2} V^{c_2} l^{d_2} + C_3 \rho^{a_3} \mu^{b_3} V^{c_3} l^{d_3} + \cdots$$

The principle of dimensional homogeneity requires that each term in the series have the same dimensions, hence

$$|F| = |\rho^a \mu^b V^c l^d| \qquad [1\cdot5]$$

where $a \cdots d$ represent any one set of the values $a_i \cdots d_i$.

The only dimensions which appear in any branch of dynamics are Mass, M, Length, L, and Time, T. Of the various terms in equation 1·5, only μ is unfamiliar, so its dimensions will first be determined. From equation 1·2

$$\mu = \frac{\tau}{\dfrac{du}{dy}}$$

and applying the dimensional homogeneity principle

$$|\,\mu\,| = \frac{|\,\tau\,|}{\left|\dfrac{du}{dy}\right|}\,\frac{|\text{Force/Area}|}{|\text{Velocity/Length}|}$$

Now $|\text{force/area}| = |\text{mass} \times \text{acceleration/area}| = ML/T^2L^2 = M/LT^2$

$$\left|\frac{\text{Velocity}}{\text{Length}}\right| = \frac{L/T}{L} = \frac{1}{T}$$

Therefore

$$|\,\mu\,| = \frac{M/LT^2}{1/T} = \frac{M}{LT}$$

All the other terms are familiar:

$$|\,F\,| = \frac{ML}{T^2}\,;\quad |\,\rho\,| = \frac{M}{L^3}\,;\quad |\,\mu\,| = \frac{M}{LT}\,;\quad |\,V\,| = \frac{L}{T}\,;\quad |\,l\,| = L$$

By substituting in equation 1·5

$$\frac{ML}{T^2} = \frac{M^a}{L^{3a}}\,\frac{M^b}{L^bT^b}\,\frac{L^c}{T^c}\,L^d$$

and by collecting terms

$$\frac{ML}{T^2} = \frac{M^{a+b}L^{c+d-3a-b}}{T^{b+c}}$$

Now if the dimensions on the two sides are equal the following is true:

$$a + b = 1;\quad c + d - 3a - b = 1;\quad b + c = 2$$

Here there are three equations with four unknowns, so that it is possible to solve for any three of the unknowns in terms of the fourth. By solving for a, c, and d, in terms of b, we easily obtain

$$a = 1 - b,\quad c = 2 - b,\quad d = 2 - b$$

so that the typical term of the series for F becomes

$$\rho^a\mu^bV^cl^d = \rho^{1-b}\mu^bV^{2-b}l^{2-b} = \rho V^2l^2\left(\frac{\mu}{\rho Vl}\right)^b$$

The relations just found between a, b, c, d will obviously hold for any set a_i, b_i, c_i, d_i; so that the series (1·4) for F becomes

$$F = \underset{i}{\Sigma}C_i\rho V^2l^2\left(\frac{\mu}{\rho Vl}\right)^{b_i} = \rho V^2l^2\underset{i}{\Sigma}C_i\left(\frac{\mu}{\rho Vl}\right)^{b_i}$$

The ratio μ/ρ occurs so frequently in fluid mechanics that it is given a name and symbol:

$$\nu = \frac{\mu}{\rho} = \text{Kinematic viscosity}$$

Furthermore the ratio $\rho Vl/\mu = Vl/\nu$, which is readily seen to be a dimensionless or pure number, is of such fundamental importance that it has been named in honor of its discoverer, Osborne Reynolds:

$$R = \frac{Vl}{\nu} \qquad \text{Reynolds number}$$

The series $\sum_i C_i R^{-b_i}$ where the values of C_i and b_i are quite arbitrary can represent any reasonable function of R, say $F(R)$. Hence equation $1\cdot4$ may be written

$$F = F(R)\rho V^2 l^2$$

It is customary to call the dimensionless function $F(R)$ a force coefficient and write it as $\tfrac{1}{2}C_F$ or $\tfrac{1}{2}C_F(R)$, the latter form being used when the dependence on Reynolds number is to be emphasized. (The reason for the $\tfrac{1}{2}$ will appear later.)

The dimensional analysis has therefore led to a very definite form for the relation between forces due to the motion of solids through fluids and the physical parameters involved. This relation is usually written

$$F = C_F \tfrac{1}{2}\rho V^2 S; \qquad C_F = C_F(R) \qquad [1\cdot6]$$

where the coefficient C_F is a pure number depending on the Reynolds number Vl/ν, and S is a characteristic area of the body in question (proportional to l^2). Actually it is found experimentally that in many cases the variation of C_F with R is negligible, so that in such cases C_F becomes a simple numerical constant.

The problem of determining and classifying experimental data for the use of the engineer is now enormously simplified. For a given shape of body at a given attitude it is only necessary to obtain experimentally the dependence of C_F on R. Then for any combination of size, speed, density, and viscosity, the force F can be calculated from equation $1\cdot6$. If C_F is a constant a single experimental value suffices to give all the required information.

In aeronautics the engineer is primarily concerned with two components of force and one moment. These are

> Lift = L = force perpendicular to the direction of motion, and usually vertical.

Drag $= D =$ force in the direction of motion, taken as positive when the force is in the downstream direction.

Pitching moment $= M =$ moment about an axis perpendicular to the direction of motion and to the lift vector, tending to raise the leading edge of the body.

The question of dimensionless coefficients corresponding to moments has not been considered but since | moment | = | force × length | it is clear that in the expression for moment analogous to equation 1·6 there must be one more length on the right side.   This length is usually chosen as the l which appears in the definition of Reynolds number.   When airplane wings or complete airplanes are under discussion, S is taken as the wing area and l is replaced by c, the mean chord of the wing.   For the three primary airplane coefficients we therefore have

$$C_L = \frac{\text{Lift}}{\frac{1}{2}\rho V^2 S} = \text{Lift coefficient}$$

$$C_D = \frac{\text{Drag}}{\frac{1}{2}\rho V^2 S} = \text{Drag coefficient}$$

$$C_M = \frac{\text{Pitching moment}}{\frac{1}{2}\rho V^2 Sc} = \text{Pitching moment coefficient}$$

$$R = \frac{Vc}{\nu} = \text{Reynolds number}$$

[1·7]

where the usual definition of Reynolds number has also been included for the sake of completeness.   It must be remembered that in general

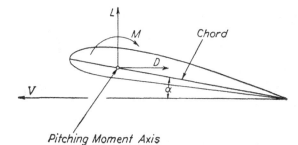

FIG. 1·4.   Conventions related to an airfoil.

all of these coefficients are functions of the shape and attitude of the body under consideration.   For a given shape of wing or airplane the attitude

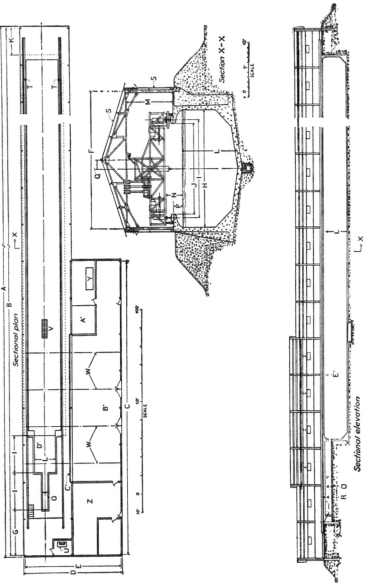

FIG. 1·5. N.A.C.A. towing basin or tank. (From *N.A.C.A. Tech. Rep. 470.*)

is generally described by the angle between the direction of motion V and a reference axis often chosen as the chord line. This angle is called the angle of attack and is denoted by α. Figure 1·4 illustrates the various quantities defined above, for an airplane wing.

1–2. Experimental Determination of Forces and Moments; Scale Effect

The most obvious procedure for the experimental determination of the forces and moments acting on a body of given shape due to its motion

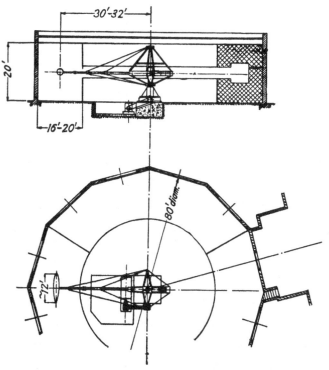

FIG. 1·6. Guggenheim Airship Institute whirling arm.

through a fluid would be to tow the body, or a geometrically similar model of it, through the fluid and measure the forces exerted on the towing members. This is actually the standard procedure in naval architecture where ship models are tested in a towing basin. In aeronautics it is primarily used in connection with seaplane float and hull characteristics. Fig. 1·5 shows a diagram of the large N.A.C.A. towing basin at Langley Field. A somewhat more convenient technique for the measurement of air forces involves the use of the whirling arm, which is

rotated at high speed while carrying a model on its end, so that the
model is essentially towed along a circular path. The whirling arm
of the Guggenheim Airship Institute at Akron* is shown in Fig. 1·6.

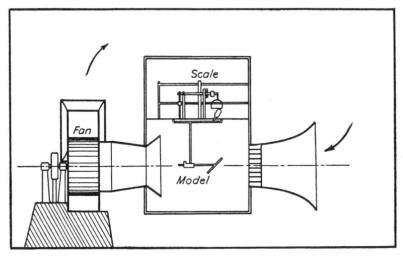

FIG. 1·7. Eiffel wind tunnel.

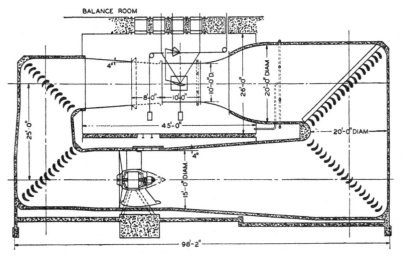

FIG. 1·8. GALCIT 10-foot wind tunnel.

  Both the above methods have obvious restrictions and difficulties
which very early led to the adoption of a quite different technique based

* Th. Troller, "The New Whirling Arm," *Jour. Aero. Sci.,* Vol. 1, October, 1934

on the elementary relativity principle already discussed. Since only the relative motion between the body and fluid is of importance the desired result is obtained if the model is held fixed and a uniform stream of fluid is driven over it. When the fluid in question is air the device for accomplishing this is called the wind tunnel. Many different types of wind tunnel have been built, three of which are illustrated in Fig. 1·7-1·9. All have the purpose of creating a very uniform stream of air moving at high velocity through a working section in which a model is

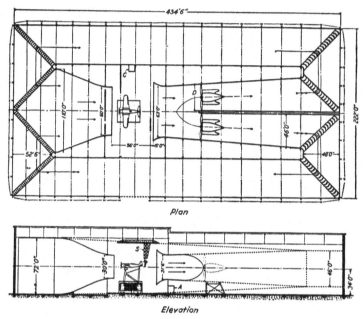

FIG. 1·9.   N.A.C.A. full-scale wind tunnel.   (From *N.A.C.A. Tech. Rep.* 459.)

mounted (Fig. 1·10). The model supports are attached to a system of balances which give the forces and moments acting on the model at various attitudes and tunnel air speeds (Fig. 1·11).   From the measured forces, the geometrical characteristics of the model (giving S and c), the observed tunnel air speed (V), and the properties of the air in the tunnel (giving ρ and μ) the appropriate dimensionless coefficients corresponding to equations 1·7 can be determined.   The resulting data are presented in tabular form or in graphs like that of Fig. 1·12 which corresponds to tests on a conventional airplane model.

It will be noted that the results correspond to a definite value of the Reynolds number. For completeness it would be necessary to have a family of such graphs covering the entire range of Reynolds numbers to be expected in the free flight of the airplane. It is almost never possible to obtain such a family, so that the question of the variation with Reynolds number of the aerodynamic coefficients becomes a vitally important one. This variation is usually referred to as *scale effect* and

Fig. 1·10. Working section of N.A.C.A. full-scale wind tunnel.
(From *N.A.C.A. Tech. Rep.* 459.)

will be discussed in detail later. For the moment it will suffice to consider it very briefly.

Some of the coefficients have only a slight scale effect, i.e., vary little with Reynolds number in the region of interest, while others may have a large scale effect. It is therefore very desirable to make the model tests at a Reynolds number as near as possible to the expected full-scale one. Remembering that $R = Vl/\nu$, we see that if the tests are made in a normal atmospheric wind tunnel the values of ν for model and full-scale conditions will be nearly the same, since ν depends only on the nature of the fluid, which is atmospheric air in both cases. This means that we should have $(Vl)_{\text{model}} \doteq (Vl)_{\text{full scale}}$. Later it will be seen that it is not possible to exceed the full-scale V of a modern high speed airplane by very much without introducing "compressibility" effects which completely alter the aerodynamic forces involved. Hence in order to get $R_{\text{model}} \doteq R_{\text{full scale}}$ the model size must be nearly that of the full scale airplane, as must the model wind velocity. Wind tunnels capable of

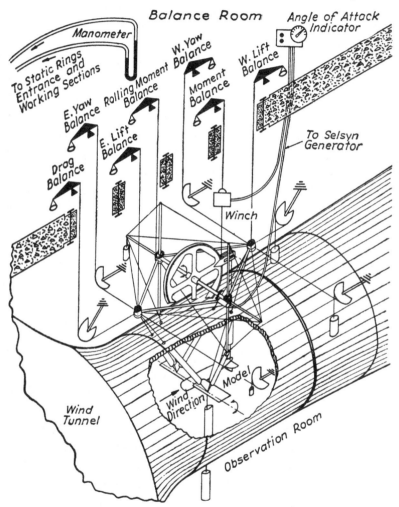

FIG. 1·11.   Wire suspension system of GALCIT wind tunnel.   (*This type of rigging is now becoming obsolescent; however, the essential features of the force resolution are shown more clearly than is possible with more complicated systems.*)

testing a full-size airplane at anything approaching full-scale velocity are enormously expensive to build and operate, and the cost of appropriate models is also extremely high. There are accordingly only a very few such tunnels in the world, all in governmental laboratories. The great bulk of the wind-tunnel testing is carried out in smaller wind tunnels at reduced Reynolds numbers, and the results extrapolated to full scale.

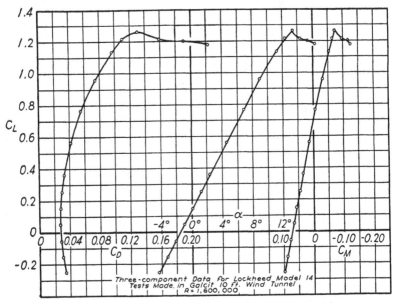

Fig. 1.12.   Aerodynamic data for typical airplane model.

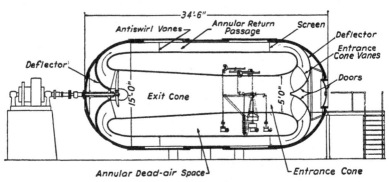

Fig. 1·13.   N.A.C.A. variable density wind tunnel.
(From *N.A.C.A. Tech. Rep.* 416.)

The variable density tunnel, the first of which (cf. Fig. 1·13) was built by the N.A.C.A. following Dr. Max Munk's design, represents a very ingenious method of getting around this difficulty. Such a tunnel is completely enclosed in a heavy steel tank which is strong enough so that the air inside may be compressed up to a pressure of 20 to 25 atmospheres. Since for air μ is approximately independent of pressure while ρ is proportional to pressure it follows that at 20 atmospheres ν has one-twentieth of its normal value at atmospheric pressure. Accordingly for a given value of Vl the Reynolds number is increased by a factor of 20 when the variable density tunnel is compressed to 20 atmospheres pressure.

It is appropriate to conclude this section on experimental methods with the numerical values for the fundamental constants of air under standard conditions of temperature and pressure:

$$\rho = \rho_0 = 0.002378 \ \frac{\text{slugs}}{\text{ft.}^3} \quad \text{at } 15° \text{ C. and 760 mm. Hg}$$

$$\nu = \nu_0 = 0.0001566 \ \frac{\text{ft.}^2}{\text{sec.}} \quad \text{at } 15° \text{ C. and 760 mm. Hg}$$

Note that

$$\rho = \frac{1}{g} \times \frac{\text{lb.-weight}}{\text{ft.}^3}$$

1–3. Aerodynamic Fluids; Streamlines and Velocity Distribution; Bernoulli's Theorem

Since some idea has been obtained of the nature of aerodynamic forces, of the methods available for their determination, and of the dimensionless coefficients which constitute the most convenient system for dealing with them, the problem must be studied more deeply. It is necessary first to consider the special characteristics of the particular fluid, air, with which we are concerned. The most important features of air for our present purposes can be crudely indicated as follows:

(a) It has small viscosity.
(b) It has small compressibility.

These statements are "crude" because they have no physical significance until the word "small" is defined by comparing it with something else. More refined statements corresponding to (a) and (b) will later be obtained; for the moment it is only necessary to make the assumption that for certain purposes air may be considered as having zero viscosity and compressibility. (The same assumption proves to be very useful for many problems involving water.) We shall accordingly, as a first approximation, define aerodynamic fluids as 'incompressible, perfect fluids," where perfect means "having zero viscosity." From equation

1·2 it follows that for a perfect fluid all shearing stresses vanish, so that, with this approximation, aerodynamic forces arise solely from normal pressures acting over the surfaces of the bodies in question.

The question therefore occurs as to whether it is possible to obtain at least a rough picture of the distribution of pressures around the surface of a body moving through an incompressible, perfect fluid. The approach to this problem turns out to be somewhat round-about and starts with the conception of "streamlines." For simplicity consider a two-dimensional steady flow. 'Two-dimensional' has been defined; by "steady" is meant that the flow conditions remain unaltered at successive instants of time. In other words, if the flow were made visible and a series of instantaneous photographs were made at different times, all of the photographs would be identical. Let us now suppose that, in a steady flow, every particle of fluid passing a fixed point could be colored and made visible, e.g., by injecting smoke into air or ink into water. Then it is clear that all of these particles would move downstream along a

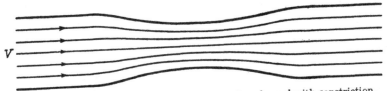

FIG. 1·14. Streamline picture for flow through a channel with constriction.
(Venturi.)

definite path which would be indicated by a color band. Such a curve is called a streamline and has the property that it is tangent to the fluid velocity at every point along its length. Another very important and obvious property is that such a streamline cannot have any end unless fluid is created or destroyed, so that streamlines either extend to infinity or form closed paths.

As a simple example consider the two-dimensional flow from left to right through a channel with a constriction (Fig. 1·14), assuming that the velocity over the cross section at the left is uniform. A number of streamlines are drawn in starting from the left with a uniform spacing, so that the same quantity of fluid is flowing through the space between each pair of neighboring streamlines. Since the velocity is everywhere tangential to the streamlines there is no flow across them, so that for the incompressible fluid the same quantity of fluid flows between two neighboring streamlines at all points along their length. From the properties of streamlines mentioned above it is obvious that the stream-line picture must have the character indicated, i.e., the streamlines are crowded closer together in the constricted than in the wider portions of the channel. Figure 1·15 represents a volume element bounded by two neighboring streamlines, two planes parallel to the paper and unit distance apart, and two faces perpendicular to the flow taken anywhere along the streamlines. The area of the left face is s_1 and the velocity across it is v_1, while s_2 and v_2 correspond to the right face. Consider the fluid which at a given instant occupies the rectangular box whose base is the face s_1 and whose altitude extends a distance v_1 upstream. After unit time the particles which were originally at the left end of the box

have just reached the surface s_1, so that all of the fluid originally in the box, and only this fluid, has crossed the face s_1. The volume of the box is s_1v_1 so that the volume of fluid entering the element through the left-hand face is just s_1v_1. Similarly the volume of fluid leaving the element in unit time through the right face is s_2v_2. Since no fluid is created or destroyed in the element and there is no flow across the sides, and since the fluid is incompressible, it follows that

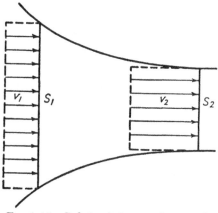

Fɪɢ. 1·15. Relation between velocity and cross-section area.

the volume of fluid entering the element in unit time through the left face is equal to the volume leaving through the right face, i.e., $s_1v_1 = s_2v_2$. Since the faces may be taken anywhere along the streamlines we conclude that

sv = constant along any pair of streamlines, where
s = streamline spacing

Hence $v \sim 1/s$ or in words: "In the streamline-flow pattern the velocity is high when the streamline spacing is narrow and low when the spacing is wide."

As a second example consider a steady flow from left to right around a circular cylinder, the flow far from the cylinder being unaffected by the latter's presence and having uniform velocity. We introduce streamlines as before, which far from the cylinder have the form of straight, parallel lines with uniform spacing (the solid lines of Fig. 1·16). In the space between the two streamlines on either side of the body which are far enough away to remain practically straight, the conditions are essentially like those of Fig. 1·14, except that the constriction arises because of a central obstruction rather than from a necking down of the walls. Accordingly the streamline picture in this region must have the character shown by the dotted lines. The relation between streamline spacing and flow velocity is obviously exactly the same as before, so that it is seen at once that the velocity is low over the front and rear of the cylinder and high over the top and bottom.

With a little experience and with the use of only geometrical feeling it is possible to draw quite successfully the streamline picture for flow of the type considered, past any quite simply shaped body. With such a streamline picture the general principle that velocity is inversely proportional to streamline spacing permits a fairly satisfactory estimate of the

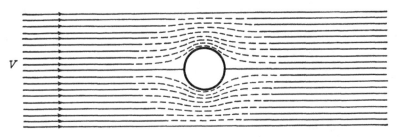

Fig. 1·16.   Streamline picture for flow around a circular cylinder.

distribution of velocity throughout the fluid and especially over the surface of the body. The problem of determining the pressure distribution over the surface would now be solved if a relation between fluid velocity and fluid pressure could be found. Such a relation was given many years ago (in 1738) by Daniel Bernoulli and constitutes the foundation of the fluid mechanics used in both aeronautics and hydraulics. In view of its fundamental importance we shall derive Bernoulli's theorem, as it is called, in its simplest form.

A very small element of volume through which the incompressible, perfect fluid is flowing with two-dimensional steady motion is shown in Fig. 1·17. The sides of the element are taken as made up of streamlines so that they are everywhere tangent to the velocity, and the size of the element is so small that these sides may be considered as straight. The two sides not shown in the diagram are taken as formed by the plane of the paper and the parallel plane unit distance away from the paper. The ends of the element are chosen perpendicular to the velocity as shown. The area of the upstream face of the element is s, that of the downstream face is $s + ds$, the pressure and velocity over the upstream face are p, v; those over the downstream face are $p + dp$, $v + dv$, and the average pressure over the sides is then $p + \frac{1}{2}dp$.

Newton's laws of motion are applied to the element in the following form: "The resultant force to the right on the element = the rate at which momentum to the right is created in the element." The force on the element arises entirely from the normal pressure forces acting over the surface since the fluid is assumed to be perfect, so that $\mu = 0$ and there are no tangential forces. The rate at which momentum is created in the element could appear in two ways: through an increase in momentum inside and through a net flow of momentum out through the boundaries. The first is zero because of the assumption of steady motion which states that the velocity and, hence, momentum at any point do not change with time. With regard to the second, the sides of the

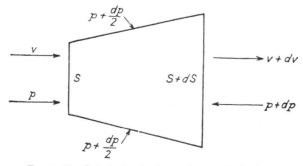

Fig. 1·17.   Infinitesimal volume element in the flow:

element have been chosen so that there is no flow across them; as a consequence no momentum can flow in or out across these boundaries. The rate at which momentum to the right is created in the element is, therefore, given by the excess of that carried out through the right end per unit time over that carried in through the left end. We proceed to calculate this excess.

As was seen in connection with Fig. 1·15, the volume of fluid entering the element per second across the face s is sv. The mass of this fluid is ρsv. Since the fluid is incompressible exactly this same mass of fluid leaves the element in unit time across the right-hand face. The momentum carried in across the left face is ρsvv while that carried out across the right face is $\rho sv(v + dv)$. Accordingly the excess of that carried out over that carried in is $\rho sv(v + dv) - \rho sv^2 = \rho svdv$. Hence we have shown that:

The rate at which momentum to the right is created in the

element = $\rho svdv$ [1·8]

We must now consider the resultant force to the right due to the pressures acting over the surface of the element. Using the positive sign for force to the right and the negative sign for that to the left we have:

The force from the left-hand face $= ps$

The force from the right-hand face $= -(p + dp)(s + ds)$

$$= -ps - pds - sdp - dpds$$

If the element becomes vanishingly small, the terms ds and dp both become small while $dpds$ becomes small so much more rapidly as to be negligible. Accordingly we drop the last term and obtain:

The resultant force from the ends of the element $= -pds - sdp$.

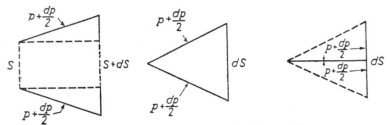

FIG. 1·18.   Pressure force from the element sides.

For the force to the right due to pressure on the sides it is clear that the resultant is identical for the three cases shown in Fig. 1·18, of which the final one obviously gives

$$(p + \tfrac{1}{2}dp)ds = pds + \tfrac{1}{2}dpds$$

By the same reasoning as that used above we see that the second term may be neglected compared to the first, so that:

The resultant force from the sides of the element $= pds$

Combining this with the force from the ends we obtain finally

The resultant force to the right acting on the element $= -sdp$ [1·9]

Equating (1·8) and (1·9) in accordance with our formulation of Newton's laws

$$\rho svdv = -sdp$$

or

$$dp + \rho vdv = 0$$

Remembering that the increments dp and dv are those which occur as we move along a streamline we may integrate along the streamline and obtain Bernoulli's theorem:

$$p + \tfrac{1}{2}\rho v^2 = B \qquad \text{Bernoulli's theorem} \qquad (1\cdot10)$$

where B is constant along any given streamline and is called the Bernoulli constant or total pressure.

Before applying Bernoulli's theorem to the problem posed at the beginning of the present section, three further remarks regarding the theorem itself should be made. The first is the fact that Bernoulli's theorem may be considered as representing a special form of the principle of conservation of energy. The term $\tfrac{1}{2}\rho v^2$ is the kinetic energy of unit volume of fluid, and it can be shown that p gives the capacity for doing work, and hence constitutes a kind of potential energy, of unit volume. The Bernoulli constant or total pressure therefore represents the total mechanical energy of unit volume of fluid, and the theorem states that this total energy remains constant as the fluid moves along a streamline. The assumption of zero viscosity is equivalent to the assumption of zero friction, so that we have been considering a system with no energy dissipation and might have expected to find just such an energy relation as Bernoulli's theorem gives.

The second remark deals with the variation of the Bernoulli constant from streamline to streamline. In a region of zero velocity the pressure p must be constant since we are not considering any gravity effects. Hence from equation $1\cdot10$ B must also be constant throughout such a region. Consider now a region throughout which a fluid is moving with the same velocity both in magnitude and direction. To an observer moving with the fluid the velocity throughout the region is zero so that again p and hence B are constant. Hence for any flow whose streamlines start in a region of constant (or zero) velocity the Bernoulli constant has the same value not only along a given streamline but throughout the entire flow. The examples of Fig. $1\cdot14$ and $1\cdot16$ have this character, as do most of the problems considered in the field of aerodynamics of the airplane.

The third remark is in the nature of a warning, namely that Bernoulli's theorem has been deduced on the assumption of a steady motion. Hence in applying it to the motion of a body through a fluid, a frame of reference *must* be adopted in which the body is at rest and the fluid is flowing past it with a steady motion.

Returning now to the original problem it is convenient to rewrite Bernoulli's theorem in the form

$$p = \text{Constant} - \tfrac{1}{2}\rho v^2 \qquad [1\cdot11]$$

where the constant may be considered as large and positive. In this form the theorem states that:

Where the velocity is large, the pressure is low;
where the velocity is small, the pressure is high.

An idea of the velocity distribution around a body has been obtained previously from a consideration of the streamline pattern so that the nature of the pressure distribution can be immediately deduced and an estimate of the resultant force obtained. In the case of the Venturi tube of Fig. 1·14 it would appear that the pressure should be high in the wide portion and low in the narrow portion, and this is usually found to be true. For the circular cylinder of Fig. 1·16 the pressure should be high at front and rear and low at the sides. Experiment also verifies this except for the high pressure at the rear. The discrepancy arises because the assumption of a perfect fluid with no viscosity breaks down in connection with the flow behind the cylinder. We shall later discuss such failures of the perfect fluid theory in considerable detail.

Figure 1·54 shows streamline patterns, velocity and pressure distributions, and resultant force components for a number of examples including the two already considered. In this connection it should be mentioned that the mathematical theory of hydrodynamics makes it possible to calculate exactly, for bodies of any shape, all the elements which have been determined very roughly by using primarily only geometrical intuition.

1–4. Consequences of Bernoulli's Theorem: Dynamic Pressure, the Pitot Tube, and Compressibility

Bernoulli's theorem gives a relation between pressure and two other quantities: $\frac{1}{2}\rho v^2$ and B which, in view of their association with p in the equation, may be given interpretations in terms of pressure. In order to find these interpretations consider a perfect fluid flow which far upstream has a uniform velocity V_0 and pressure p_0, so that B is constant throughout the fluid. This flow passes over a symmetrical body A with rounded front as shown in Fig. 1·19. The streamlines over the upper half of the body are indicated. For the moment, neglect the manometer and the small tubes leading from the body to it; then the flow is completely symmetrical so that the central streamline follows the symmetry axis from far upstream to the body at point (1). Here it must divide and pass along the top and bottom surface of the body until the downstream point (3) on the symmetry axis is reached. Here the two portions of the streamline recombine and the streamline proceeds downstream again along the axis of symmetry. At the point (1) where this streamline divides into two and turns through 90°, it is clear that the velocity V_1 is zero. Such a point where the velocity vanishes is called a "stagnation point." It is possible to design the shape of the body so that at some point along its surface—say at (2)—the velocity V_2 is exactly equal to the "free-stream" velocity V_0. Now applying Bernoulli's theorem we have, since B = constant,

$$p_0 + \tfrac{1}{2}\rho V_0^2 = p_1 + \tfrac{1}{2}\rho V_1^2 = p_2 + \tfrac{1}{2}\rho V_2^2 = B$$

with the conditions given above

$$V_1 = 0, \qquad V_2 = V_0$$

Hence

$$\left.\begin{array}{l} p_1 = p_0 + \tfrac{1}{2}\rho V_0^2 \\[4pt] p_2 = p_0 \end{array}\right\} \qquad [1 \cdot 12]$$

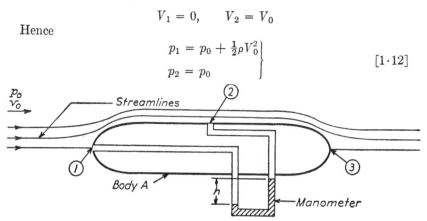

Fig. 1·19.   Idealized Pitot-static tube and manometer.

The first equation shows that the pressure at the stagnation point exceeds the free-stream pressure by the amount $\tfrac{1}{2}\rho V_0^2$. This additional pressure is obviously associated with the impact corresponding to the stopping of the fluid particles originally moving with velocity V_0. It is accordingly called the "impact pressure" or more commonly the "dynamic pressure" and is often denoted by the symbol "q." The definition of dynamic pressure is then:

$$q = \tfrac{1}{2}\rho V^2 = \text{Dynamic pressure} = \text{pressure produced by bring-}$$

ing to rest particles of fluid originally moving with velocity V. $[1 \cdot 13]$

The fundamental importance of this quantity furnishes the explanation for the factor $\tfrac{1}{2}$ which was introduced without any apparent reason in the aerodynamic coefficients of equations 1·7.

Since the second term of Bernoulli's equation 1·10 has been given the name "dynamic pressure" it is desirable to give the term p a somewhat more explicit title. It is therefore referred to as the "static pressure" in order to avoid any confusion. The Bernoulli constant B is then called the "total pressure."

At the surface of the specially shaped body of Fig. 1·19 point (1) is subjected to the total pressure $p_1 = p_0 + \frac{1}{2}\rho V_0^2$, while point (2) experiences the static pressure p_0. Small orifices at (1) and (2) permit these pressures to be transmitted through tubes to a pressure measuring instrument or manometer outside of the body as shown schematically in the figure. Such a manometer will record the difference in pressure $p_1 - p_2$. But from equations 1·12

$$p_1 - p_2 = \Delta p = \tfrac{1}{2}\rho V_0^2$$

so that the head, h, exhibited by the manometer is proportional to the dynamic pressure of the flow.

A tube with a single orifice like (1) is called a total pressure tube; one with an orifice like (2), a static pressure tube; and one with both

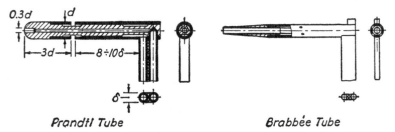

Prandtl Tube   Brabbée Tube

Fig. 1·20.   Prandtl and Brabbée Pitot-static tubes.

orifices, a Pitot-static tube, or more commonly, although incorrectly, a Pitot tube. Since a Pitot-static tube, when combined with a pressure gage, gives the dynamic pressure it can be used to determine the air speed V_0 when the density is known. Conventional air-speed meters used on aircraft consist of such a Pitot-static tube connected to a pressure gage calibrated to indicate velocity for standard air density. Figure 1·20 shows two types of Pitot-static tube which have been widely used.

In addition to furnishing the basis for convenient experimental methods of velocity determination, Bernoulli's equation makes it possible to obtain an estimate of certain effects of the air's compressibility. To investigate this question consider the compression which occurs at the stagnation point of a body moving with velocity V_0 through air which far away is at rest and has the normal atmospheric pressure $p_0 = 14.7$ lb./in.^{2} To apply Bernoulli's equation we must consider the body at rest in a flow which far away has pressure p_0 and uniform velocity

$-V_0$, so that we have a steady motion and B is constant throughout the fluid. Applying Bernoulli's equation to the region far from the body and to the stagnation point on the body

$$p_0 + \tfrac{1}{2}\rho V_0^2 = p_s$$

where p_s = pressure at the stagnation point where $V = 0$. Now, writing the difference between the normal atmospheric pressure and that at the stagnation point as Δp, we have

$$\Delta p = p_s - p_0 = \tfrac{1}{2}\rho V_0^2 = \text{Pressure change due to motion}$$

Consider the percentage change in pressure referred to the normal atmospheric pressure p_0

$$100\,\frac{\Delta p}{p_0} = 100\,\frac{\tfrac{1}{2}\rho V_0^2}{p_0} = \text{Per cent change in pressure due to motion}$$

Assuming standard sea level conditions $\rho = 0.002378$ slugs/cu. ft. and $p_0 = 2116$ lb./sq. ft. Hence using consistent units so that V_0 is in feet per second

$$100\,\frac{\Delta p}{p_0} = \left(\frac{V_0}{133}\right)^2 \quad \text{or} \quad V_0 = 133\,\sqrt{100\,\frac{\Delta p}{p_0}}$$

Hence a velocity of 133 ft./sec. or 91 m.p.h. corresponds to only a 1 per cent change in pressure. If the compression occurred isothermally, i.e., at constant temperature, the percentage change in density would be the same as that in pressure. Actually the compression occurs adiabatically, i.e., with temperature changes, so that the percentage change in density is only about 0.7 as large as that in pressure. If the calculations are carried out accurately it appears that the change in density usually is negligible for speeds up to about one-half the velocity of sound, i.e., up to 350 to 400 m.p.h. As the speed increases above this value aerodynamic characteristics are altered, at first slowly and then more and more rapidly until the speed of sound (1130 ft./sec. under standard conditions) is reached, when sudden discontinuities occur.

It therefore appears that for normal airplane characteristics compressibility plays no role, and the air may be considered to be as incompressible as water. The use of the terms compression and rarefaction, which is very common in connection with airplane lift and drag, is therefore completely unjustified. The pressures which cause these forces have nothing to do with compression, but are associated with velocity differences in accordance with the Bernoulli theorem. This is probably the most fundamental conception of aerodynamics for the aeronautical engineer.

1–5. Curved Flow; Circulation; Theory of Lift; Airfoils

In developing the so-called circulation theory of lift it is convenient to introduce a very simple type of flow, namely, the steady, two-dimensional motion of a perfect fluid in concentric circles, the Bernoulli constant having the same value throughout the fluid. For such a flow the streamlines will obviously constitute a family of concentric circles. For purposes of visualization it is convenient to consider one of these streamlines as coinciding with a circular cylinder so that the flow is taken as occurring around the outside of such a cylinder (Fig. 1·21). Consider the equilibrium of a very small element at a distance r from

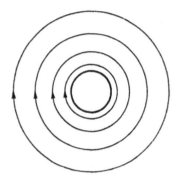

FIG. 1·21. Streamlines for concentric circle flow.

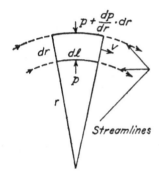

FIG. 1·22. Volume element for the flow of Fig. 1·21.

the center of curvature (Fig. 1·22). The centrifugal force on the fluid in the element must be just balanced by the resultant force due to pressures over the surface. Considering the element to have unit thickness perpendicular to the diagram the volume is $dldr$ and the centrifugal force is $\rho dldr \cdot v^2/r$. The resultant pressure force acting towards the center of curvature * is

$$\left(p + \frac{dp}{dr}\,dr\right)dl - pdl = \frac{dp}{dr}\,drdl$$

Equating these forces and canceling $drdl$ we obtain

$$\frac{dp}{dr} = \frac{\rho v^2}{r}$$

---

* A more accurate analysis taking into account the pressures over the radial faces as shown for Figs. 1·17 and 1·18, leads to exactly the same results.

Now from equation 1·10, or more directly from the equation preceding 1·10, we have, since B is taken as constant throughout the fluid,

$$dp = -\rho v dv$$

Hence

$$-\rho v \frac{dv}{dr} = \frac{\rho v^2}{r}$$

or

$$\frac{dv}{v} = -\frac{dr}{r}$$

and integrating

$$\log v = -\log r + \text{constant}$$

Therefore

$$\log v + \log r = \text{constant}$$

$$\log vr = \text{constant}$$

$$v = \frac{\text{constant}}{r} \qquad\qquad [1·14]$$

Hence it has been proved that in such a flow as that which has been considered the velocity is inversely proportional to the radius of curvature of the streamline, i.e., to the distance from the center of the flow. The streamlines of Fig. 1·21 have been drawn so that their spacing indicates this decrease in velocity with increasing radius.

If such a flow existed around a point in the fluid itself rather than around a cylinder, then as $r \rightarrow 0$, $v \rightarrow \infty$. The point $r = 0$ has therefore very peculiar properties and is called a singular point or in hydrodynamics, a vortex. The circulatory flow around such a vortex is often called a vortex flow and is of fundamental importance in aerodynamics. Since infinite velocities cannot exist in reality, the vortex point must either be a fictitious one lying outside the fluid as in the example of Fig. 1·21, or if it lies in the fluid the conditions in the neighborhood of $r = 0$ must be different from those corresponding to equation 1·14. When vortices do occur in a fluid they are called free vortices and it is found that the effect of viscosity (which was neglected in deriving equation 1·14) is such as to cause the portion of fluid near $r = 0$ to rotate like a solid body with $v \backsim r$. Outside of this central core, which is called the vortex core, is a transition region, and outside of this the fluid rotates in accordance with equation 1·14.

By applying Bernoulli's equation with constant total pressure to such a free vortex flow it is seen that the static pressure decreases rapidly towards the center as is shown in the diagram of Fig. 1·23. If the vortex has considerable intensity and the core has a small diameter, extremely high suctions can occur in the core. The tornado and waterspout are striking examples of the occurrence in nature of vortex flows with just these properties. Later, it will be seen that every airplane in flight also trails behind it similar vortex flows whose effects, although less catastrophic than those of the examples just cited, are still very unpleasant.

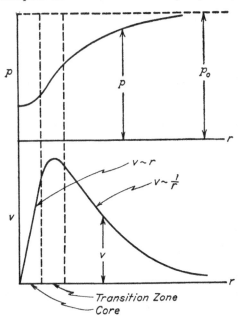

FIG. 1·23. Pressure and velocity distribution in a vortex flow.

In order to generalize some of the characteristics found above to less specialized flows, it is necessary to introduce the conception of circulation. This has a very precise mathematical definition which will first be stated for the benefit of readers with mathematical training. A detailed explanation will then be attempted for those without such training. Consider a given closed path drawn through a region filled with moving fluid. Let v be the fluid velocity at any point along the path, dl be an element of length along the path, and θ the angle between v and dl (Fig. 1·24). Then the circulation, written as Γ, around the given path or contour is the line integral of $v \cos \theta$ around the path, or using the conventional mathematical notation

$$\Gamma = \oint v \cos \theta \, dl \qquad [1·15]$$

It is easy to see what this means by recalling that integration is essentially nothing more than a process of summation. Let us approximate to our closed path by a polygon of n sides, the lengths of which are $\Delta l_1, \Delta l_2, \cdots \Delta l_n$. Let the fluid velocity at the center of each side be $v_1, v_2, \cdots v_n$, and the angles between the corresponding v's and l's be $\theta_1, \theta_2 \cdots \theta_n$ (cf. Figs. 1·24 and 1·25). Now consider the sum

$$v_1 \cos \theta_1 \, \Delta l_1 + v_2 \cos \theta_2 \, \Delta l_2 + \cdots v_n \cos \theta_n \, \Delta l_n = \sum_i v_i \cos \theta_i \, \Delta l_i$$

obtained by multiplying each length by the tangential component of the velocity at its center and adding all the products together. Now let the number of subdivisions increase while their lengths decrease; then the "line integral" of equation 1·15 is just what is obtained if the limit $n \rightarrow \infty$ is approached while the length of each $\Delta l \rightarrow dl \rightarrow 0$. The word "line" modifying "integral" means that the whole procedure and the integral itself are only defined if a particular path or line is specified along which the summation or integration is carried out.

In order to make this a little clearer let us calculate, as an example, the circulation around one of the circular streamlines of the flow of

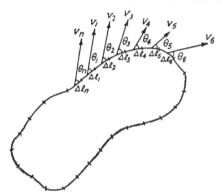

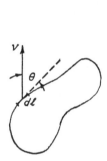

Fig. 1·24. Velocity
at a line element.

Fig. 1·25.   Elementary components of
circulation.

Fig. 1·21. Remember that the velocity is everywhere tangential to such a path and its magnitude at any point is given by equation 1·14 which is rewritten

$$v = \frac{A}{r} \qquad [1\cdot16]$$

giving the constant of equation 1·14 the symbol A. Since the velocity is everywhere tangential to the chosen path (cf. Fig. 1·26), θ is everywhere 0 and $\cos \theta = 1$. If we let $d\varphi$ be the central angle subtended by the path element dl, then

$$dl = rd\varphi$$

Hence equation 1·15 becomes

$$\Gamma = \oint v \cdot 1 \cdot rd\varphi = \oint \frac{A}{r} rd\varphi = \oint A d\varphi$$

introducing the expression (1·16) for v. The symbol crossing the integral sign indicates the adoption of the convention that the integration is to be carried out in a clockwise direction around the contour, i.e., from an arbitrary value of φ, say φ_1, to $\varphi_1 + 2\pi$. Since A is constant, independent of φ

$$\Gamma = A \oint d\varphi = A \int_{\varphi_1}^{\varphi_1 + 2\pi} d\varphi = A\left[(\varphi_1 + 2\pi) - \varphi_1\right] = 2\pi A \quad [1\cdot17]$$

In this especially simple case the result could have been obtained without any integration for, since the tangential component of velocity

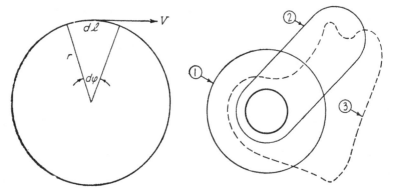

FIG. 1·26. Diagram for circulation with concentric circle flow.

FIG. 1·27. Alternative paths having the same circulation.

$v \cos \theta = v = A/r$ is the same for every point on the path, Γ reduces to the simple product of this velocity times the length of path or

$$\Gamma = \left(\frac{A}{r}\right) 2\pi r = 2\pi A$$

A very remarkable fact appears in the result given by equation $1\cdot17$ which is that the magnitude of the circulation is independent of the distance r, i.e., Γ is the same no matter which of the circular streamlines outside the body of Fig. 1·21 we choose for the path of integration. A still more remarkable result can be rigorously proved, although the proof lies outside the limits of the present discussion; namely, for the flow in question the circulation Γ has the same value for every closed path which encloses the cylinder just once, e.g., paths 1, 2, and 3 of Fig. 1·27. The final generalization which is proved in texts on hydrodynamic theory, and which gives a fundamental importance to the conception of circulation may be roughly stated in the form: " In a perfect fluid with constant total pressure the circulation is identical around every simple * closed path enclosing a given set of solid bodies or vortices."

This fundamental theorem has as a consequence the fact that the phrase " the circulation around a body " has a unique and well-defined meaning, independent of the choice which may be made of the particular path used for calculating the value of the circulation.

One more conception is required before the above considerations can be applied to the problem of lift which is the objective of the present section. This is the principle of superposition which states that it is possible to build up complex flows by the superposition or addition of a number of simpler flows. In making this addition it must be remembered that the fundamental quantities to be added are velocities, so that the addition must be a vectorial, and not a scalar one. The problem is exactly the same as that of finding the resultant of several forces by adding the various components vectorially. In considering the flow superposition, the velocity vectors of the component flows must be added vectorially at every point to obtain the final resultant or composite flow. As an example, the superposition of a flow with uniform horizontal velocity and another with uniform and equal vertical velocity gives a resultant flow with uniform velocity whose direction is inclined at 45° to the horizontal and whose magnitude is $\sqrt{2}$ times that of the two component flows. Figure 1·28 illustrates the vectorial addition which must be carried out at every point of the flow when such a superposition is made.

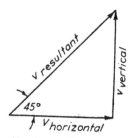

Fig. 1·28. Vectorial addition of component flow velocities.

Figure 1·29 shows velocity vectors and the corresponding streamline patterns for three flows around a circular cylinder. The first is a rectilinear flow with velocity V from left to right which has zero circulation. The fact that the circulation vanishes can be seen by choosing a circular path of integration concentric with the circle and by remembering that for every point above the horizontal line of symmetry there is a symmetrical point below. For the point on top the tangential velocity is in the direction of integration (clockwise), while for that on the bottom it has the same magnitude but is opposite to the direction of integration. Hence $v \cos \theta \, dl$ has equal and opposite values for the two symmetrical points so that the corresponding terms cancel when the integration (or

* The word "simple" means primarily that none of the paths compared has a loop so that one or more of the bodies or vortices is passed around more than once in making a single circuit of the path.

summation) is carried out. The entire integration is made up of such pairs of equal and opposite terms so that the final Γ is zero. The second flow is a pure circulatory flow with circular streamlines. It has a circulation, Γ, but no net rectilinear flow. The third is the compound flow

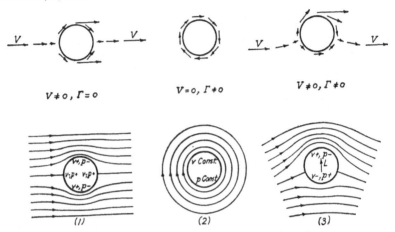

FIG. 1·29. Perfect fluid flows around a circular cylinder.

obtained by superimposing the other two. It has both a rectilinear velocity V and a circulation Γ. In order to avoid confusion only a few velocity vectors are indicated in Fig. 1·29, but enough appear to show the effect of the vector summation involved in the superposition. The streamline pattern can be deduced from the velocity vector diagram by remembering that the streamlines are the curves which are everywhere tangent to the velocity. It should be noted that the existence of circulation as in (3) does not imply that particles of fluid actually rotate completely around the body forming closed streamlines. If the circulation is large enough compared to V and the size of the body, such closed streamlines may exist; however, usually combined flows of practical importance do not have closed streamlines, but have the general character shown in the diagram. Figures 1·30 and 1·31*

Courtesy of McGraw-Hill Book Co.

FIG. 1·30. Flow around circular cylinder without circulation.

* Reproduced from Prandtl-Tietjens, "Applied Hydro- and Aero-Mechanics," McGraw-Hill Book Co., 1934.

present photographs obtained at Göttingen of flows without and with circulation around a circular cylinder.   The streamlines have been made visible and their similarity to those shown diagrammatically in Fig. 1·29 is very striking.

From the results of the earlier discussions a very good estimate of the distribution of velocity and pressure over the surface of the cylinder can be obtained.   In the streamline diagrams of Fig. 1·29 the relative magnitudes of velocity and pressure have been indicated by using + to signify "large" and − to signify "small."   The only relations used in obtaining the indicated values are those giving the streamline spacing-

Courtesy of McGraw-Hill Book Co.

FIG. 1·31.   Flow around circular cylinder with circulation.

velocity connection, and the Bernoulli theorem.   By studying the diagrams with the view of finding the resultant pressure force it can be seen that for the first two flows everything is symmetrical about both a horizontal and a vertical axis through the center of the cylinder.   Hence all the pressures balance one another and there is no resultant force. In the third diagram a lack of symmetry appears for the first time, but only about the horizontal axis.   The figure is still symmetrical about the vertical axis so that the pressures to the right are balanced by equal pressures to the left on the other side of the symmetry axis.   Hence there is no resultant force to left or right, i.e., no drag force.   However, the pressures below are high and those above are low so that there is a resultant vertical force which is perpendicular to V and hence appears as a lift as indicated on the diagram.

The results obtained in this very elementary fashion for the especially simple case of a circular cylinder are proved by hydrodynamic theory to have a much wider range of applicability.   In fact, for the

steady, two-dimensional flow of a perfect fluid about a cylindrical body
of any shape whatever, the following statements may be rigorously
proved to be correct:*

(a) The drag is zero.
(b) In the absence of circulation around the body the lift is zero.
(c) If there is a circulation of magnitude Γ around the body and if
the rectilinear flow velocity past the body is V then a lift exists,
the magnitude per unit length perpendicular to the flow being
given by

$$L' = \rho V \Gamma \qquad \text{Kutta-Joukowsky law} \qquad [1\cdot18]$$

(a) and (b) together constitute what is known as D'Alembert's
paradox and were for many years after their discovery in 1744 considered as indicating the uselessness of any theoretical approach to the
practical problems of fluid mechanics. It is now realized that they represent a very important first approximation in a wide field of aerodynamic phenomena. The third statement (c) furnishes the foundation for
the entire modern conception of flight. This famous law can be quali-
tatively visualized very easily as follows: Consider a body moving
horizontally to the left through a stationary fluid. Then, relative to
the body, the fluid is flowing from left to right. Suppose that the body
has a clockwise circulation when looking in such a direction that the fluid
comes from the left and moves off to the right. Then above the body
the rectilinear flow velocity and the circulation flow velocity are in the
same direction so that they are added. Below the body the circulation-
flow velocity is in the opposite direction to the rectilinear flow so that the
circulation-flow velocity is subtracted from the rectilinear-flow velocity.
Hence above there is a high velocity and a resulting low pressure
while below there is a low velocity and hence a high pressure. The
combination of pressures produces an upward force or lift which is proportional to the rectilinear velocity and the circulation. This physical
picture was first given by Lord Rayleigh, but the precise mathematical
relationships were worked out in the early 1900's by the German mathe-
matician Kutta and the Russian scientist Joukowsky, whose names have
been attached to the final law, equation 1·18.

Figure 1·32 shows schematically two types of flow around a wing
section or airfoil at a low and a high angle of attack. The first, (a),
represents flow with zero circulation, while the second, (b), corresponds
to the same flow with non-vanishing and specified circulation. The
characteristic feature of the flow without circulation, (a), is that the

* Flows with velocity discontinuities are here not considered.

fluid flows around the pointed trailing edge to the downstream stagnation point on the upper surface.   If the trailing edge is sharp, as by definition it is for an airfoil, its radius of curvature is zero and hence, from the concentric circle flow discussion, the velocity around it is infinite.   In order to eliminate this infinite velocity Kutta in 1902 assumed that a circulation is set up around the airfoil of strength just sufficient to move the rear stagnation point downstream to the trailing edge.   The flow would then stream smoothly off both upper and lower surfaces leaving the airfoil just at the trailing edge and eliminating the infinite velocity.   Although Kutta could not give a physical explanation for this phenomenon experiment shows that his assumption accurately

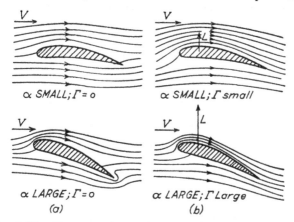

Fig. 1·32.   Flows around an airfoil without and with circulation.

corresponds to reality, and some years later Prandtl gave a complete description which will be discussed in section 1–7.

Several important elements of the flow around airfoils can be inferred from the diagrams of Fig. 1·32.   Probably the most important is the relation between Γ and the angle of attack α.   It is clear that, for the flow without circulation, the rear stagnation point moves further and further from the trailing edge as α increases.   Hence the strength of the circulation required to keep the stagnation point at the trailing edge increases with α.   Theory and experiment agree in the result that over the range of practically used angles of attack Γ is proportional to α_a, where α_a is called the absolute angle of attack and represents the angle of attack measured from the zero-lift attitude of the airfoil where $\Gamma = 0$.

Since circulation is a sum or integral of terms of the form (velocity $\times$ length) it is seen that the dimensions of Γ are $|$ velocity $| \times |$ length $|$. The only velocity which appears in Fig. 1·32 is the flow velocity V, and the most convenient length is the airfoil chord c. Accordingly we write

$$\Gamma \sim Vc\alpha_a$$

or, using $\frac{1}{2}a_0$ to denote the constant of proportionality

$$\Gamma = \tfrac{1}{2}a_0Vc\alpha_a \qquad [1\cdot19]$$

From equation $1\cdot18$ this gives for the lift per unit length of the cylindrical airfoil (perpendicular to the plane of the diagram)

$$L' = a_0\alpha_a\tfrac{1}{2}\rho V^2 c \qquad [1\cdot20]$$

In equation $1\cdot7$ we have already given an expression for the total lift on a wing:

$$L = C_L\tfrac{1}{2}\rho V^2 S$$

where S is the wing area. If we use this same form for the lift of unit span of a wing of chord c, then $S = c \times 1$ and we have

$$L' = C_L\tfrac{1}{2}\rho V^2 c \qquad [1\cdot21]$$

Comparing equations $1\cdot20$ and $1\cdot21$ we obtain

$$C_L = a_0\alpha_a \qquad [1\cdot22]$$

so that the lift coefficient is proportional to the angle of attack (cf. Fig. $1\cdot12$). If α_a is measured in radians (1 radian = 57.3°) then hydrodynamic theory gives

$$a_0 = 2\pi \qquad \text{(Theory)} \qquad [1\cdot23]$$

for conventional thin airfoils. For practically all modern airfoil sections experimental results for the normal range of flight Reynolds numbers may be expressed in the form

$$a_0 = 2\pi\eta \quad \text{where} \quad \eta \doteq 0.90 \qquad \text{(Experiment)} \qquad [1\cdot24]$$

The correction factor η is usually called the " airfoil efficiency factor."

The second important element which appears from Fig. $1\cdot32$ is the pressure distribution over the airfoil surface. If, as is customary, the pressure is measured from the undisturbed static pressure in the rectilinear flow far from the airfoil, then at the stagnation point the pressure will have its maximum value of $q = \frac{1}{2}\rho V^2$, i.e., the full dynamic pressure. Referring to the figure it is seen that over the entire lower surface the pressure will be high, although not as high as q since the fluid is moving, even if relatively slowly. Over the leading edge, particularly on the upper surface, the velocity is higher than V so that the pressure is lower than the reference pressure and we have a suction. Proceeding downstream along the upper surface the intensity of this suction decreases until at the trailing edge it reaches nearly the undisturbed static pressure again.   The intensity of the suction peak near the leading edge increases rapidly as α increases.   Figure 1·33 shows typical airfoil pressure distribution diagrams in which all pressures are made dimensionless by dividing by q, pressures greater than the free-stream static pressure are called positive, and pressures lower than this reference

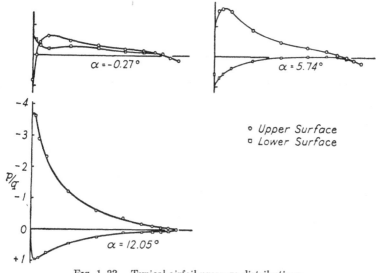

Fig. 1·33.   Typical airfoil pressure distributions.
(From *N.A.C.A. Tech. Rep.* 614, Fig. 3.)

pressure are called negative.   The lift is approximately proportional to the area inside the curve.

1–6. The Wing of Finite Span; Prandtl Wing Theory

The most important features of the two-dimensional flow of a perfect aerodynamic fluid about cylindrical bodies perpendicular to the flow have now been discussed.   These considerations have led to an explanation of lift but not of drag.   The latter force component first appears when the discussion is extended to the case of wings of finite span.   Here it is obvious that the flow is not identical in all planes perpendicular to the span, so that the generalization really consists in the extension from two- to three-dimensional flow.   The most satisfactory treatment of this problem of the three-dimensional flow over a wing of finite span was given by Prandtl in 1918, although the basic ideas were first discussed by him in 1911, and had been realized still earlier by Lanchester. Prandtl's treatment is intimately connected with certain properties of vortices, which must therefore be stated before the wing problem itself can be treated.

A vortex was defined in connection with two-dimensional motions as the point at the center of a circulation flow where the velocity became infinite. In three dimensions the theorem of the constancy of circulation holds exactly as in two dimensions, i.e., in a flow with circulation the circulation is the same around every path enclosing a given body or vortex. Hence if the length of the path around a vortex is shrunk to zero the velocity becomes infinite at the vortex. The curve connecting the points where this occurs is called a vortex filament. The strength of the vortex at any point along the filament is defined as the magnitude of the circulation around the filament. The vortex laws, which are proved in the hydrodynamic theory of perfect fluids, and which are important here, may be stated as follows:

(a) A vortex filament cannot have an end but must either continue to infinity or form a closed path. This is the same property as was found in Section 1-3 in connection with streamlines.

(b) The strength of such a vortex filament is constant along its length.

(c) Vortices in a fluid always remain attached to the same particles of fluid.

These are called Helmholtz' vortex theorems and were first given by Helmholtz in 1858. A related theorem which we shall require later was given at about the same time by Thomson (Lord Kelvin) and is called Thomson's theorem:

(d) The circulation around any path which is always made up of the same fluid particles is independent of the time.

All these theorems are based on the fact that in a perfect fluid no tangential forces can act, and hence the angular velocity of a fluid particle can never change since no couple can be exerted on it. We shall not attempt any further explanation of the theorems but shall proceed with their application to the problem of the finite wing.

Earlier it was seen that in two dimensions the existence of lift was associated with the presence of circulation and it can be shown that the same relation must hold in three dimensions. Accepting this result it follows that around every section of a finite lifting wing there must exist a circulation, Γ. As far as the lift is concerned conditions would be unaltered if the solid wing were replaced by a vortex having the same strength, Γ, since in both cases the lift per unit span is given by the Kutta-Joukowsky law

$$L' = \rho V \Gamma$$

The vortex filament which replaces the wing without altering the lift, is called by Prandtl the lifting line. In order to indicate that the vortex itself has no physical reality in the fluid but is in effect bound to the inside of the wing it is referred to as a "bound vortex" cf. Fig. 1·34).

Prandtl's fundamental conception was to realize that Helmholtz' vortex laws applied to such bound vortices precisely as if they were ordinary vortices having a physical existence in the fluid. Therefore the theorems (a) and (b) indicate that the bound vortex cannot end at the wing tips where the lift falls to zero. The only possibility is that they continue laterally out of the wing and become actual physical

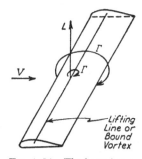

Fig. 1·34. The bound vortex replacing a wing.

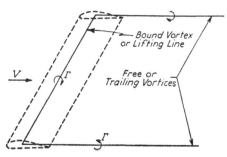

Fig. 1·35. Prandtl's horseshoe vortex system.

vortices in the fluid. In this condition they are called "free vortices" to distinguish them from "bound vortices" which are attached to the wing. Theorem (c) requires that these free vortices as they are formed at the wing tips trail off downstream along the streamlines made up of the particles of fluid which strike the tips. They are therefore often referred to as trailing vortices. If the motion is a steady one, the trailing vortices continue downstream to infinity. In order to simplify the picture and make calculations possible Prandtl assumes that the trailing vortices follow the undisturbed streamlines, which are straight lines parallel to V drawn through the wing tips, rather than following the actual streamlines from the tips. This approximation is found to lead to negligible errors in most practically important cases. The simplified picture of the complete system of vortices is now as indicated in Fig. 1·35 and is often called the horseshoe vortex pattern. As appears on the diagram, the strength of the trailing vortices must be the same as that of the bound vortex and the sense of rotation is from outside → over → in as indicated.

Each of the trailing vortices has associated with it a circulatory motion like that shown in Fig. 1·21 for flow around a cylinder. Hence a view from in front would show a streamline pattern and velocity distribution like those indicated in Fig. 1·36 where the top half of the diagram shows the streamlines in a vertical plane through the lifting line, and the lower half shows the velocity at the wing which is directed vertically downward and is called the "downwash velocity, w." The distribution of w over the span can be estimated from the fact that the velocity associated with each trailing vortex follows the law already deduced for two-dimensional flow (equation 1·14), namely $v \sim \Gamma/r$ where r is the distance from the vortex in question. The downwash

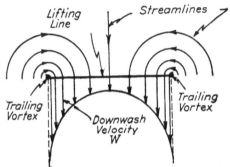

FIG. 1·36. Streamlines and downwash associated with trailing vortices.

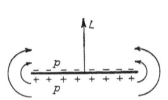

FIG. 1·37. Lateral flow arising from pressure distribution.

velocity w is obtained by adding the velocities associated with the two trailing vortices.

Two justifications for the above picture can be given at this point before proceeding further with the analysis. First, consider the distribution of pressures which must exist above and below a wing which is furnishing lift. The front view of Fig. 1·37 shows the high pressures below $(+)$ and the low pressures above $(-)$ which combine to give the lift. At the tips there must be a lateral flow from the high pressure around to the low pressure. This lateral flow around the tips is that shown in Fig. 1·36 associated with the trailing vortices. Second, the trailing vortices can be made visible by smoke and are sometimes observed in flight, when planes are maneuvering in atmospheric conditions which are such as to make water condensation occur in the reduced-

pressure region of the trailing-vortex cores.   These vortices then appear
as filaments of mist or fog trailing behind the wing tips.

Returning to the problem, consider the conditions which exist at an
airfoil section of the wing lying between the two tips.   At this section
there exists the rectilinear-flow velocity, V, which, if the span were
infinite so that the trailing vortices were infinitely far away, would be
the only outside velocity.   The flow would then be two-dimensional
and the results of the previous section would be applicable.   However,
because of the finite span there actually exists, in addition to the velocity
V, a finite downwash velocity w at the airfoil section, so that the resultant
velocity is V_{res} as indicated in Fig. 1·38.   Prandtl now assumes that
the conditions at the airfoil section are the same as if the latter were a
section of a cylindrical wing in
a two-dimensional flow with
rectilinear velocity V_{res}.   In
other words, as far as an observer located on the section
is concerned he experiences only
the rectilinear flow, V_{res}, without knowing whether it actually
comes from infinity or not.
From this it follows that the
lift per unit length of span at
the section is given by $\rho V_{res} \Gamma$
and is perpendicular to V_{res},

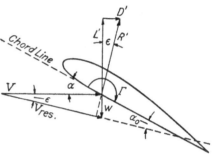

FIG. 1·38.   Velocity and force triangles at a
wing section.

since the results of the two-dimensional theory are carried over to
the present case.   This lift is the only force acting on the section and is
indicated by the vector R' (representing resultant force per unit length
of span) on the diagram.   For an observer considering the forces on
the wing as a whole the lift will be defined as the component of the
resultant force perpendicular to the true velocity at infinity, V, i.e., to
the flight path of the wing if the fluid is considered as at rest and the
wing as moving through it.   Similarly the drag will be the component
of force in the direction of V.   Hence the resultant force R' furnishes
both a lift component L' and a drag component D' as shown in Fig.
1·38.   From the figure we see that the velocity and force triangles are
similar, so that

$$\frac{L'}{R'} = \frac{V}{V_{res}} = \cos \epsilon; \qquad \frac{D'}{L'} = \frac{w}{V} = \tan \epsilon$$

where ϵ is the " downwash angle " between V and V_{res}.   It is assumed in Prandtl's theory and verified by experiment that $w \ll V$, so that ϵ is a small angle and to a first approximation

$$\cos \epsilon = 1; \quad \tan \epsilon = \sin \epsilon = \epsilon$$

Accordingly

$$L' = R'; \quad \frac{D'}{L'} = \frac{w}{V} \qquad [1 \cdot 25]$$

Consideration of the angle of attack shows that the apparent angle of attack of the section (sometimes called the geometrical angle of attack) is the angle between the chord line and the velocity at infinity V. This appears in Fig. 1·38 as α, which is the usual designation. However, to the observer on the airfoil section the angle of attack, which is the angle actually determining the circulation, is that between the chord line and V_{res}. This is shown as α_0 on the diagram and is usually called the "effective angle of attack." In other words the airfoil section at the apparent angle of attack α experiences exactly the same forces as if it were a section of an infinite span wing at an angle of attack α_0 in a two-dimensional flow.

These results may be summarized as follows: The effect of finite span is to introduce trailing vortices which produce a downwash at the wing. This downwash rotates the effective velocity vector which decreases the effective angle of attack and rotates the resultant force vector backwards, so that a drag component is introduced. These two effects are connected with the downwash velocity by the relations:

$$\alpha_0 = \alpha - \frac{w}{V}; \quad \frac{D'_i}{L'} = \frac{w}{V} \qquad [1 \cdot 26]$$

The subscript $()_i$ has been attached to the drag symbol because of the fact that this drag force is usually called "induced drag." The downwash velocity w is also often referred to as "induced velocity." The reason for this terminology is that the relation between downwash velocity and trailing vorticity is analogous to that between a magnetic field strength $(\sim w)$ and the electric current in a wire $(\sim \Gamma)$ inducing the field. The numerical computations involved in Prandtl's theory are identical with calculations developed long ago in the theory of electromagnetic induction, so that the adjective "induced" has been taken from the electrical field and introduced into aerodynamics. It should be pointed out and emphasized that induced drag is not a resistance associated with friction since a perfect, frictionless fluid has been considered throughout. The energy which must be supplied to overcome this drag in the case of a moving airplane is not therefore immediately dissipated in the form of heat, but is left behind in the fluid as the kinetic energy associated with the circulatory motion around the trailing vortices.

The discussion up to this point has related to conditions at a particular airfoil section of a finite wing. In order to determine the characteristics of the wing as a whole it is necessary to carry out a summation or integration of these elementary effects over the entire span, but difficulties present themselves immediately. For, referring to Fig. 1·36, consider an element very close to one wing tip. Here, since the induced velocity is inversely proportional to the distance from the trailing vortex, $w \rightarrow \infty$ as the tip is approached. But a fundamental assumption of the theory was that w is very small. The complete resolution of this contradiction was only achieved by Prandtl's group after years of effort, although it now seems very simple. Instead of considering a single lifting line with its associated trailing vortices, let us visualize a number

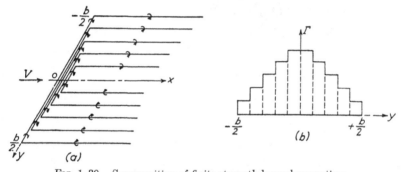

FIG. 1·39.   Superposition of finite strength horseshoe vortices.

of such lifting lines of different lengths superimposed on one another as indicated in Fig. 1·39 (a) where the x axis is an axis of symmetry parallel to V and the y axis lies along the span. The distribution of trailing vortices appears in (a) and the distribution of circulation along the span in (b). The distribution of circulation follows from the fact that circulations may be superimposed and the effects added algebraically. Since for every section along the span the Kutta-Joukowsky equation, $L' = \rho V \Gamma$, holds the distribution of lift along the span is exactly proportional to the distribution of Γ. The superposition principle can obviously be carried to the limit in which an infinite number of horseshoe vortices each of infinitesimal strength is added together in the manner suggested by Fig. 1·39. The result indicated in Fig. 1·40 will be a trailing vortex sheet rather than a number of discreet vortices, and a continuous distribution of circulation (and hence of lift) over the span instead of a stepwise distribution. No single vortex will have a finite strength so that there are no infinite values of the downwash, and the difficulties mentioned above disappear.

With this very flexible picture it is possible to determine a distribution of the elementary horseshoe vortices which corresponds to any desired lift distribution over the span. The distribution of circulation or bound vorticity determines that of the trailing vortices and hence of the

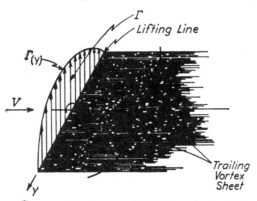

downwash velocity, so that the lift and downwash distributions are very closely related. The downwash in turn determines the effective angle of attack in accordance with equation 1·26 and this, in connection with the airfoil section geometry, specifies the circulation at any point of the span. The working out of this complex set of interrelationships to give the characteristics of a given wing involves very lengthy and elaborate calculations which will later be discussed briefly.

Fig. 1·40. Continuous distribution of circulation and trailing vorticity.

However, there is one particular type of lift distribution, discovered by Prandtl, for which everything is very simple. This is the so-called "elliptic lift distribution." If y represents distance along the span from the center of the wing (cf. Fig. 1·39), b the wing span, L' the lift per unit length along the span at an arbitrary point y, and L_0' the value of L' at the center of the wing, then the elliptic lift distribution is expressed analytically by

$$L' = L_0' \sqrt{1 - \left(\frac{y}{b/2}\right)^2} \qquad [1·27]$$

or

$$\frac{L'^2}{L_0^2} + \frac{y^2}{(b/2)^2} = 1$$

The reason for the name is apparent since a plot of L' vs. y has the shape of an ellipse.

The special properties associated with the elliptic lift distribution are

(a) The downwash is constant along the span.

(b) For a given total lift, span, and velocity V the induced drag has its lowest possible value.

(c) The elliptic distribution theoretically occurs for an untwisted wing of elliptical plan form.

(d) Lift distributions over untwisted wings of conventional tapered plan forms are theoretically and experimentally found to approximate closely to the elliptical shape.

The simplifications accompanying these special properties are so great that practically all preliminary calculations for normal wings are based on the assumption of this optimum elliptical lift distribution. In particular the property (a) implies that the velocity triangle of Fig. 1·38 is identical for every airfoil section along the span. In other words the ratio $D_i'/L' = w/V$ (equation 1·26) has the same value at every wing element. When the forces on all these elements are added together to furnish the total wing force, it follows that the total induced drag and lift have the same relation, i.e.,

$$\frac{D_i}{L} = \frac{w}{V}$$

or dividing numerator and denominator of the left side by $\frac{1}{2}\rho V^2 S$ to give force coefficients

$$\frac{C_{D_i}}{C_L} = \frac{w}{V}$$

If the "induction" calculations are carried out for this case of elliptical lift distribution a very simple result is obtained for the downwash. This result is so basic to most aerodynamic calculations that it belongs with the Bernoulli theorem and the Kutta-Joukowsky law in the category of the few formulas which the aerodynamicist must memorize. It is

$$\frac{w}{V} = \frac{C_L}{\pi \mathcal{R}}$$

where

$$\mathcal{R} = \text{Aspect ratio} = \frac{b^2}{S}$$

For a wing with rectangular plan form the aspect ratio is simply the ratio of span to chord.

Collecting the above results we obtain the primary relations given by the simple Prandtl wing theory:

$$\left.\begin{aligned}
\frac{w}{V} &= \frac{C_L}{\pi \mathcal{R}} = \text{Constant over the span} \\[2mm]
\alpha_0 &= \alpha - \frac{C_L}{\pi \mathcal{R}} \\[2mm]
C_{D_i} &= \frac{C_L^2}{\pi \mathcal{R}}
\end{aligned}\right\} \quad \begin{aligned} &\text{Elliptic} \\ &\text{lift} \\ &\text{distribution} \end{aligned} \quad [1\cdot 28]$$

An alternative form of the last relation which is very convenient for many purposes is obtained by multiplying both sides of the equation by qS where q is the dynamic pressure $\frac{1}{2}\rho V^2$:

$$C_{D_i} qS = \frac{C_L^2 qS}{\pi \mathcal{R}} \cdot \frac{qS}{qS} = \frac{(C_L qS)^2}{\pi(b^2/S)qS} = \frac{L^2}{\pi q b^2}$$

Therefore

$$D_i = \frac{L^2}{\pi q b^2} \qquad\qquad [1\cdot 29]$$

This form is especially useful when calculations of total force are desired. In equations $1\cdot 28$ and $1\cdot 29$ the quantities C_L, C_{D_i}, L, and D_i refer to

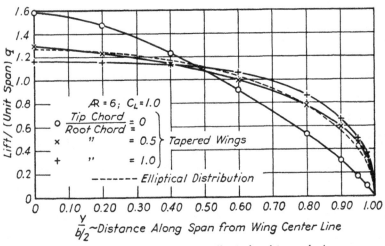

FIG. $1\cdot 41$.   Lift distributions over elliptical and tapered wings.

the over-all characteristics of the wing rather than to the characteristics of any particular section along the span.

Figure $1\cdot 41$ presents the results of calculations giving the distribution of lift over the semi-span of several untwisted wings of aspect ratio 6, having taper ratios covering the range of values normally employed on contemporary airplanes. The curves are plotted from tabular data pre-

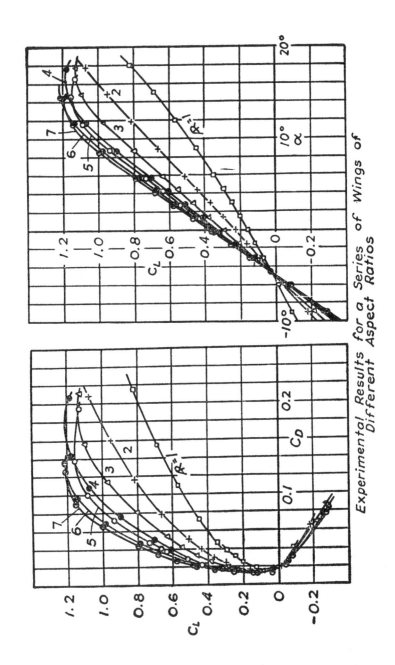

Experimental Results for a Series of Wings of
Different Aspect Ratios

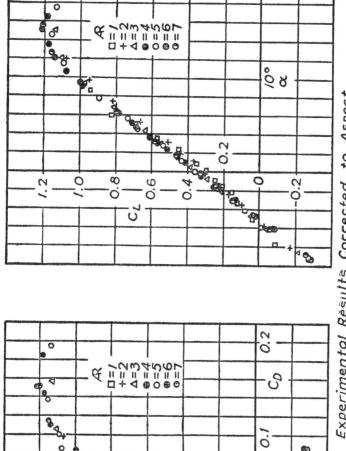

Experimental Results Corrected to Aspect Ratio 5

Fig. 1·42. Experimental verification of Prandtl wing theory.
(From *N.A.C.A. Tech. Rep.* 116.)

sented by Anderson.* The elliptical lift distribution is also included for comparison, and the various lift distributions are seen to approximate fairly closely to the elliptical shape, in agreement with property (d) mentioned above following equation (1·27).

Before discussing the further extensions of the theory to cases involving other lift distributions it will be desirable to consider some of the consequences of the results obtained thus far. Probably the most important is that giving the dependence of total wing drag on plan form or, more precisely, on aspect ratio. The fundamental assumption in this connection is that the induced drag of a wing represents that portion of the total drag which depends on wing plan form, any remaining drag depending on the shape of the airfoil profiles which make up the wing. This assumption implies that if two untwisted rectangular wings, 1 and 2, having the same airfoil section and chord, but differing in span or aspect ratio, are tested under similar conditions their drag coefficients will differ only by the amount of the difference in induced drag coefficients:

$$C_{D_1} - C_{D_2} = \frac{C_L^2}{\pi \mathcal{R}_1} - \frac{C_L^2}{\pi \mathcal{R}_2} = \frac{C_L^2}{\pi}\left(\frac{1}{\mathcal{R}_1} - \frac{1}{\mathcal{R}_2}\right) \qquad [1·30]$$

Similarly the apparent angles of attack at which the two wings will attain a given value of C_L will differ by the downwash angle difference for the two wings:

$$\alpha_1 - \alpha_2 = \frac{C_L}{\pi}\left(\frac{1}{\mathcal{R}_1} - \frac{1}{\mathcal{R}_2}\right) \qquad [1·31]$$

The relations given by equations 1·30 and 1·31 furnish a very powerful tool whereby tests need be made at only one aspect ratio, the characteristics for any other aspect ratio being obtained by simple calculations from those at the original one. These relations also permit a relatively easy but powerful experimental check to be made of the theory. Such a check was carried out at Göttingen under Prandtl's direction soon after the theory was developed. A series of wings identical in all respects except aspect ratio was tested in the wind tunnel. The results, shown in Fig. 1·42, exhibit first the great differences in wing characteristics for the various aspect ratios, and second the remarkable agreement when all of the results are converted to an aspect ratio of 5 using equations 1·30 and 1·31. When it is remembered that the theory is based on the replacement of a wing by a lifting line with zero chord, the fact that the predictions of the theory are accurately substantiated by experiment for aspect ratios as small as 2 is very astonishing.

* Raymond F. Anderson, "Determination of the Characteristics of Tapered Wings," *N.A.C.A. Tech. Rep.* 572 (1936).

Equations 1·30 and 1·31 assume an especially useful form if the second of the two wings to be compared is taken as having infinite aspect ratio. The subscript ( )₀ is applied to the C_D and α for this wing to indicate that the corresponding aspect ratio is infinite. Then dropping the subscript ( )₁

$$\left. \begin{array}{l} C_D = C_{D_0} + \dfrac{C_L^2}{\pi R} \\[2em] \alpha = \alpha_0 + \dfrac{C_L}{\pi R} \end{array} \right\} \qquad [1\cdot32]$$

The drag coefficient C_{D_0} corresponding to infinite aspect ratio is called the " profile drag coefficient." The angle of attack α_0 for the same case has already been discussed as the effective angle of attack. Both α_0 and C_{D_0} depend only on airfoil profile shape and are independent of wing plan form. In the past it was often customary to present airfoil characteristics corresponding to some standard aspect ratio (usually 5 or 6) and then to convert these data to any particular aspect ratio desired using equations 1·30 and 1·31. In recent times the much simpler procedure has been widely adopted of presenting infinite aspect ratio characteristics $C_{D_0}(C_L)$ and $\alpha_0(C_L)$ and converting these directly to any aspect ratio using equations 1·32.

It will be noticed that in the preceding sentence $\alpha_0(C_L)$ was written rather than $C_L(\alpha_0)$. This was done in order to emphasize the fact that the fundamental variable in the majority of problems involving wings is most conveniently taken as C_L. One is likely to think of the geometrical variable, angle of attack, as the independent variable and all others as dependent. However the fundamental position of C_L in the Prandtl theory makes it much more satisfactory to give this variable the basic role and to consider the others as depending on it. In other words, it is preferable to think in terms of setting a C_L and finding to what angle of attack the wing must be rotated, rather than in terms of setting an α and calculating the C_L to be expected. We shall see later that this is even more applicable when we consider the actual flight of an airplane.

One result of the Prandtl induction theory which has not yet been considered and which will be stated without proof is: "The relation between pitching moment coefficient C_M and C_L is independent of the aspect ratio." It is obvious that this is not true of the relation between C_M and α. This fact furnishes an additional reason for the choice of C_L rather than α as the independent variable.

Figure 1·43 presents typical infinite aspect ratio results for a conventional airfoil section, presented in a form which has been widely

adopted.   The symbol $C_{M_{0.25}}$ means that the moment coefficients presented are referred to a moment axis located on the chord line 25 per cent
of the chord length aft of the leading edge.   If we consider only the

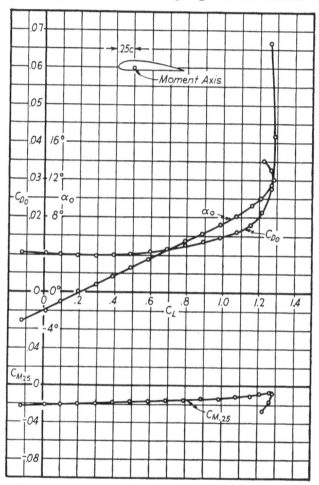

Fig. 1·43.   Typical infinite aspect ratio wing characteristics.
(From *tests at the GALCIT*; $R = 1,200,000$.)

region between zero lift and $C_L = 1.0$, which corresponds to the normal
flying range, we see that the curves exhibit certain very simple characteristics:

(a) The C_L vs. α_0 curve is practically a straight line.
(b) The C_M vs. C_L curve is also practically a straight line.

(c) C_{D_0} is very roughly constant. (This is much less accurate than a or b.)

These characteristics will be taken up in more detail in the section on *Airfoil Characteristics*. Here it is necessary only to point out that the linear relation between C_L and α_0 has already been discussed in section 1·5 where the symbol a_0 was chosen for the slope of the C_L vs. α_0 curve.

If the typical infinite aspect ratio characteristics (a) and (c) above are introduced into equations 1·32, it is possible to draw certain conclusions regarding the characteristics of wings with finite aspect ratio.

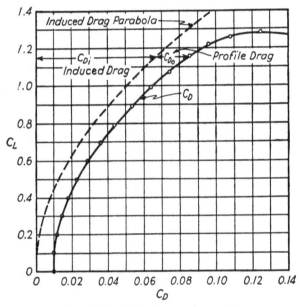

FIG. 1·44. Polar diagram for a wing.

Considering first the drag coefficient, it appears that a wing polar (Fig. 1·44), as the plot of C_L vs. C_D is called, has approximately the shape of a parabola for $0 < C_L < 1$. This first approximation of a parabolic polar will be the starting point for much of the later performance analysis.

Turning now to the angle of attack, it is convenient to introduce the notation $\alpha_{L_0} = \alpha$ for $C_L = 0$

$$\alpha_a = \alpha - \alpha_{L_0} \qquad [1·33]$$

Note that, in accordance with equations 1·32, at $C_L = 0$ the geometrical and effective angle of attack are equal, $\alpha = \alpha_0$, so that α_{L_0} refers equally to α and α_0. Now we have seen that theory and experiment agree in giving a linear relation between C_L and α_0, the constant of proportionality having been written as a_0, so that

$$C_L = a_0(\alpha_0 - \alpha_{L_0}) \qquad [1\cdot34]$$

In view of the linear relation between α, α_0, and C_L given by equation $1\cdot32$ a similar relation would be expected to hold between C_L and α. If the constant of proportionality in this case is written as "a" the anticipated relation would have the form

$$C_L = a(\alpha - \alpha_{L_0}) \qquad [1\cdot35]$$

It is now necessary to verify this relation and at the same time find the connection between a and a_0. By subtracting α_{L_0} from both sides of the second equation of $1\cdot32$

$$\alpha - \alpha_{L_0} = (\alpha_0 - \alpha_{L_0}) + \frac{C_L}{\pi\!R}$$

From equation $1\cdot34$

$$(\alpha_0 - \alpha_{L_0}) = \frac{C_L}{a_0}$$

and substituting this in the equation just above

$$\alpha - \alpha_{L_0} = \frac{C_L}{a_0} + \frac{C_L}{\pi\!R} = \frac{C_L}{a_0}\left(1 + \frac{a_0}{\pi\!R}\right)$$

Hence

$$C_L = \frac{a_0}{1 + \dfrac{a_0}{\pi\!R}}\,(\alpha - \alpha_{L_0})$$

which is exactly of the form of equation $1\cdot35$ if

$$a = \frac{a_0}{1 + \dfrac{a_0}{\pi\!R}} \qquad [1\cdot36]$$

The simple Prandtl theory for the case of elliptical lift distribution therefore gives a definite relation between the "lift curve slopes" for infinite and finite aspect ratio. The relation in equation $1\cdot36$ will appear later very importantly in the *Airplane Stability* analysis.

It only remains in concluding this section to outline the modifications to the above results which are introduced when the lift distribution is not elliptical. Although the theoretical calculations are much more complicated in this case because of the fact that the downwash is no longer constant across the span, the final results for untwisted wings appear in a

quite simple form. It is only necessary to modify equations 1·32 as
follows:

$$C_D = C_{D_0} + \frac{C_L^2}{\pi \mathcal{R}} (1 + \delta) \; \Big\rbrace$$

$$\alpha = \alpha_0 + \frac{C_L}{\pi \mathcal{R}} (1 + \tau) \; \Big\rbrace \quad \begin{array}{c} \text{Non-elliptic lift} \\ \text{distribution} \end{array} \qquad [1\cdot 37]$$

where δ and τ are small correction factors depending on the deviation
of the lift distribution from the ideal elliptical form.   Figure 1·45 shows
values of the induction factors δ and τ as calculated by Glauert for

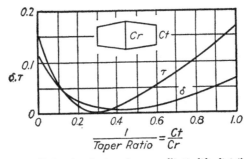

FIG. 1·45.   Induction factors for non-elliptic lift distribution.

straight-tapered, untwisted wings of aspect ratio about 6. It appears
that for normal taper ratios, C_r/C_t, the correction for induced drag is so
small as to be negligible, while that for angle of attack may become large
enough to require consideration for nearly rectangular wings ($C_t/C_r \doteq 1$).

1–7. Effects of Viscosity at Large Reynolds Numbers; The Boundary Layer

All the theoretical considerations thus far have been based on the
assumption of perfect fluids, i.e., viscosity has been entirely neglected.
Even with this extreme over-simplification the theory has explained a
group of phenomena of tremendous importance.   The lift of airplane
wings and the associated induced drag have not only been physically
clarified, but quantitative predictions agreeing remarkably well with
experiment have even been made possible.   Three types of phenomena
have, however, already been encountered which the theory has been
helpless to elucidate.   These are

   (a) The deviation of the flow pattern from that predicted at the
       rear of bluff bodies like circular cylinders.

(b) The limitation to the values of C_L which can be attained with airfoils, and the deviations from the predicted airfoil characteristics in the maximum lift region (cf. Fig. 1·12, 1·43).

(c) The existence of profile drag (cf. Fig. 1·44).

The physical mechanism by which circulation develops around a wing is also not yet clear. All of these points represent failures of the theory, associated with its basic assumption of perfect fluids, and all are satisfactorily explained by the famous boundary layer theory introduced by Prandtl in 1904 to describe the influence of viscosity under large Reynolds number flow conditions.

The starting point of Prandtl's considerations is the empirically justified "condition of no slip." This states that whenever a real fluid, no matter how small its viscosity, flows over a solid surface, however smooth, the layer of fluid immediately adjacent to the surface sticks to the surface without any slip whatever.* All of our perfect fluid con-

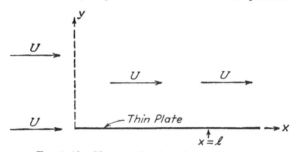

FIG. 1·46. Notation for flow along a flat plate.

siderations have been based on the assumption of zero shearing stress so that the theoretical flows, heretofore considered, slide over solid surfaces without any retardation. In order to investigate the implications of the no-slip condition consider a two-dimensional, steady flow, with uniform velocity and hence constant static pressure, over the surface of a smooth plate set parallel to the flow. A coordinate system attached to the plate is chosen as shown in Fig. 1·46, and the undisturbed velocity far from the plate is denoted by U. The plate is taken as infinitely thin so that in the case of perfect fluid flow it coincides with one of the undisturbed streamlines and produces no disturbance whatever, since the fluid adjacent to the surface slips over the latter with the velocity U.

* The only exception to this statement which has been found experimentally occurs in highly rarefied gases when the mean free path of the molecules becomes large compared to the dimensions of the solid body in question. Such cases never occur in practical aeronautical problems.

Figure 1·47 contains on the left the velocity profile at some point $x = l$ for this perfect fluid flow. On the right are drawn a corresponding series of profiles for real fluids with different values of μ. All the profiles on the right have the common features that

(a) far from the surface ($y \to \infty$) the local velocity $u = U$, and
(b) at the surface ($y = 0$) $u = 0$.

If the viscosity is small it is physically apparent that the disturbing effect of the plate will extend only a relatively short distance from the plate, while if the viscosity is large the layers of fluid much further away

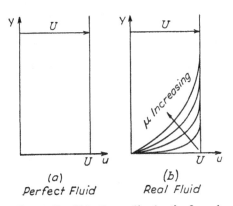

(a)
Perfect Fluid

(b)
Real Fluid

FIG. 1·47. Velocity profiles for the flow of
Fig. 1·46.

FIG. 1·48. Boundary
layer notation.

will be slowed down by the dragging action of the plate. We denote the distance from the wall at which the dragging effect of the plate has become negligible by δ (cf. Fig. 1·48), so that

$$u \doteq U \text{ at } y = \delta \qquad [1{\cdot}38]$$

Prandtl's boundary layer assumption may be very crudely stated as follows: "When a fluid has very small viscosity, $\mu \to 0$, the thickness of the fluid layer affected by viscosity is very small, $\delta \to 0$. This layer, $0 \leq y \leq \delta$, is called the boundary layer." The statement in this form, although it represents the essential elements of the theory, is actually meaningless since "small" is not compared to anything. A more correct, if less transparent, statement is the following: "If l represents distance downstream along the surface from the leading edge of a body in a flow with undisturbed velocity U, and δ is the boundary layer thickness at the point l, then if $R_l = Ul/\nu \gg 1$ then $\delta/l \ll 1$." In this form the comparisons have been given a precise meaning since they are made in the form of statements that dimensionless ratios have values very large or very small compared to 1. Prandtl's analysis applied to the flat plate flow of Fig. 1·46 actually leads to the result that

$$\frac{\delta}{l} \sim \frac{1}{\sqrt{R_l}} \quad \text{where} \quad R_l = \frac{Ul}{\nu} \qquad [1\cdot39]$$

Equation 1·39 leads to the conclusion that if R becomes very large δ/l becomes very small, so that the flow over the plate with its very thin boundary layer becomes practically identical to that given by a perfect fluid analysis. Let us investigate the behavior of the skin friction under these conditions. For this purpose consider Fig. 1·47(b) using the notation given in Fig. 1·48. As $\mu \to 0$ the velocity profile near the wall becomes fuller and fuller so that because of the no-slip condition the slope of the profile at the wall, $\tan \theta$, becomes smaller and smaller and also approaches zero. It was seen earlier in connection with Fig. 1·3 that the velocity gradient $du/dy = 1/\tan \theta$. Hence as $\mu \to 0$, $(du/dy)_0 \to \infty$. Here the subscript $(\ )_0$ denotes conditions at the surface $y = 0$. From equation 1·2

$$\tau_0 = \mu \left(\frac{du}{dy} \right)_0$$

which in the limit $\mu \to 0$ becomes

$$\tau_0 = 0 \cdot \infty$$

This is an indeterminate form which is shown by Prandtl to be proportional to $1/\sqrt{R_l}$. Hence even though R_l becomes very large, a finite skin friction remains which does not vanish until $R_l = \infty$, corresponding to a true perfect fluid. Hence for real fluids with very small viscosity, i.e., flows with very large Reynolds number, the general flow pattern closely approaches that of a perfect fluid, but the shearing stress at the wall remains finite as long as the viscosity does not actually vanish. This finite value of shearing stress means that even at enormous Reynolds numbers there is a definite skin friction and hence a drag force.

The results outlined above can often be carried over with only minor modifications to the case of flow around a body with curved surfaces such as an airfoil or fuselage. An explanation of phenomenon (c) of the first paragraph of this section has thus been furnished. For it now appears that for large Reynolds numbers the effect of viscosity may be such as to leave the main perfect fluid flow unaltered so that the airfoil characteristics already deduced are practically unaffected, while a finite skin friction is introduced. This skin friction is, in fact, the profile drag which was completely unexplained by the earlier perfect fluid theory.

We shall return to a more detailed discussion of skin friction in a later chapter and proceed here with more general considerations.

The explanation of phenomena (a) and (b) rests upon a different aspect of the boundary layer picture. Consider the boundary layer along the surface of a circular cylinder in a rectilinear flow where the conditions are like those shown in Fig. 1·29 (1). Selected boundary layer profiles for this case are shown in Fig. 1·49 with the y coordinate scale enormously expanded. In this figure the coordinate system is curvilinear, the x axis lying along the surface of the cylinder and the y coordinate being perpendicular to it. U represents the velocity in the perfect fluid flow outside the boundary layer at any point x along the surface, and U_0 is the free-stream velocity far from the cylinder.

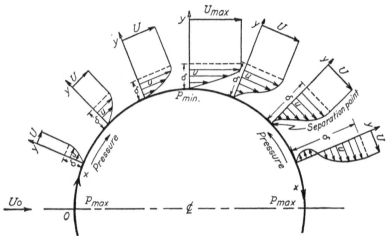

FIG. 1·49. Boundary layer along a circular cylinder.

In accordance with the discussion in section 1–3, U increases from zero at the forward stagnation point, $x = 0$, to its maximum value at the top of the cylinder, and then decreases again as one goes downstream along the rear face. Using Bernoulli's theorem it follows that the pressure first decreases in going downstream from $x = 0$ and then increases again. It is one of the important conclusions of the boundary layer theory that the pressure does not change in going perpendicularly to the flow across the thin boundary layer, so that the pressure across any boundary layer cross section is constant and equal to the value computed by the Bernoulli theorem just outside the boundary layer where viscosity has no effect and the theorem may be applied.

The characteristics of the boundary layer are quite different in the regions of decreasing pressure on the front face and of increasing pressure on the rear face, i.e., the regions of accelerated and decelerated flow. In the first region the velocity profiles are very similar to those which occur in the flat plate flow (cf. Fig. 1·48). The only difference is that now there is a pressure gradient which tends to accelerate all the fluid particles so that the velocity near the wall will be a little higher than if there were no pressure gradient. In other words the boundary layer will be somewhat thinner and the profile a little fuller in the region of accelerated flow than in the case of uniform flow. As soon as the minimum pressure point is passed in the downstream direction, however, conditions become quite different. From this point on the pressure gradient gives a force opposing the velocity so that the fluid particles in the boundary layer are decelerated instead of accelerated. This is, of course, also true of the fluid just outside of the boundary layer. For this outer portion of the fluid the kinetic energy which has been attained at the minimum pressure point, $\frac{1}{2}\rho U_{\max}^2$, is just sufficient to carry the fluid against the pressure gradient around to the rear stagnation point before the velocity drops to zero. However, in the boundary layer at the minimum pressure point the kinetic energy is less than $\frac{1}{2}\rho U_{\max}^2$ since friction has been retarding the flow and consuming kinetic energy which appears as heat through the action of viscosity. Hence the pressure gradient will slow these particles down and bring them to rest before the rear stagnation point is reached. This will obviously occur first next to the wall where the velocity was lowest at the minimum

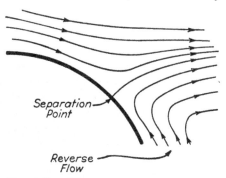

Separation Point

Reverse Flow

Fig. 1·50. Flow in the neighborhood of a separation point.

pressure point. Accordingly in the region of decelerating flow the effect of the pressure gradient is to increase the boundary layer thickness and fundamentally alter its shape by greatly slowing down the velocity near the wall. As is indicated in Fig. 1·49 a "separation point" is eventually reached where the velocity profile has a vertical tangent at the wall and where the fluid particles a little distance out have been brought to rest. Immediately downstream from this point these layers of fluid will obviously be driven backwards, or upstream, by the pressure gradient. Hence a reverse flow occurs near the surface and the main flow leaves or separates from the surface at the separation point as is indicated by the streamline pattern of Fig. 1·50. It is apparent that downstream from such a separation point the main flow has an entirely different character than that which would be expected from the earlier perfect fluid considerations. Figure 1·51 shows schematically the general characteristics of the flow around a circular cylinder where the

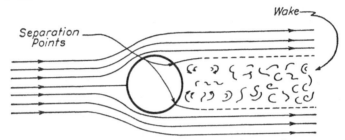

FIG. 1·51. Wake behind a cylinder with separation.

separation points occur shortly behind the minimum pressure points. The reverse flow behind the separation points is very unstable and breaks up into a large number of eddies. At moderate Reynolds numbers these eddies are arranged in a definite pattern known as the Kármán vortex street, an example of which is shown in Fig. 1·52.* At higher Reynolds

Courtesy of McGraw-Hill Book Co.

FIG. 1·52. Kármán vortex street behind a circular cylinder.

numbers (above, say, 100,000) the flow is completely irregular and eddying or turbulent. In either case the region occupied by the eddying motion is called the "wake." The existence of such a wake means that the main flow streamlines fail to close in behind the body so that

* Reproduced from Prandtl-Tietjens, *loc. cit.*

the stagnation pressure is not developed over the rear surface. This destroys the fore and aft symmetry of the flow pattern and hence of the pressure distribution so that a large pressure drag occurs.

It is now possible to distinguish between streamline and bluff bodies. The former are bodies which are so shaped that the pressure increase downstream from the minimum pressure point is so gradual that separation does not occur at least before the rear of the body is practically reached. The boundary layer is then everywhere extremely thin and the flow and pressure distribution are nearly identical with those which would be expected in the case of a perfect fluid flow. The drag of such bodies is very small, arising almost entirely from skin friction. With bluff bodies, on the other hand, the pressure rise is rapid enough so that separation occurs well ahead of the rear of the body, with the resulting formation of a wake which seriously alters the perfect fluid flow pattern and leads to a pressure drag. Such bodies always have a high drag of which the skin friction is only a small part.

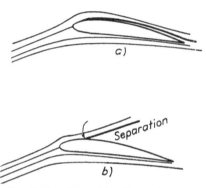

Fig. 1·53.   Alternative flows about an airfoil near the stall.

An explanation of phenomenon (a) of the first paragraph has thus been given. That for (b) is very similar and rests upon the fact that a given body may fall into the "streamline" category at one attitude and into the "bluff" category at another. This is the case for airfoils which at low and moderate angles of attack behave like streamline bodies (Fig. 1·53a) and, apart from a small frictional profile drag, exhibit characteristics which are essentially those predicted by perfect fluid theory. For many airfoils this state of affairs continues as α increases until a critical angle of attack is reached at which a separation point suddenly appears on the upper surface near the leading edge (cf. Fig. 1·53b). The flow then suddenly changes its character from the streamline to the bluff body type, a large pressure drag is developed, the circulation and lift drop off, and the airfoil is said to have stalled. The angle of attack at which this occurs is sometimes called the "burble point." This rather idealized description is not always applicable, the stall frequently occurring over a small but finite range of α during which the separation point rapidly moves forward without actually jumping to the leading edge. In such cases the drag increase is less precipitous and the peak of the C_L vs. α curve is rounded instead of having a sharp maximum. The essential physical processes, although more gradual, are still the same.

Phenomena (a) to (c) of the first paragraph have now been explained. Figure 1·54 gives a schematic summary of some of the more important

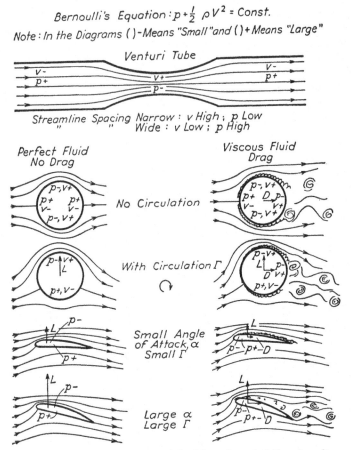

FIG. 1·54. Summary of perfect and real fluid flow characteristics; two-dimensional motion.

flow phenomena in this connection. It only remains in concluding this chapter to complete the physical picture by describing the actual origin and development of the circulation around an airfoil.

It will be remembered that in general the flow without circulation about an airfoil involves flow around the sharp trailing edge. When an airfoil is set in motion in a stationary fluid this is the type of flow which actually occurs in the first instant as has been repeatedly demonstrated by visual flow observations. The flow around the sharp corner leads to the formation of a vortex as shown in Fig. 1·55.* This is associated

Courtesy of McGraw-Hill Book Co.

Fig. 1·55. Vortex created by flow around a sharp corner.

with the fact that the fluid involved in the formation of the vortex has been in the boundary layer and hence must be considered as a viscous rather than a perfect fluid. As the motion proceeds this vortex is de-

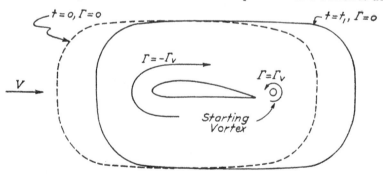

Fig. 1·56. Development of circulation around an airfoil.

tached from the airfoil surface and remains in the fluid behind as a free vortex. At this stage in the proceedings consider a contour surrounding the airfoil and some distance away as indicated in Fig. 1·56. This con-

* Reproduced from Prandtl-Tietjens, *loc. cit.*

tour is attached to a definite set of fluid particles. We adopt the point of
view of an observer attached to the airfoil so that we have a rectilinear
flow from infinity which suddenly starts from rest. Initially $(t = 0)$
everything was at rest so the circulation around the contour was zero.
At the later time $(t = t_1)$, when the trailing edge vortex has been de-
tached, the contour has moved downstream but is still far from the airfoil so that the fluid particles composing it have not been contaminated
by the action of viscosity. In other words, this contour may still be
considered as lying in a perfect fluid so that the Thomson theorem of the
constancy of circulation (cf. section 1-6) is applicable. Hence the circulation around the contour remains zero at the time t_1. The circulation
around any contour is the algebraic sum of all the circulations enclosed
by the contour. Now we have at t_1 a circulation, say Γ_v, around the

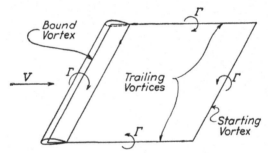

FIG. 1·57.   Complete vortex pattern (idealized).

free vortex which has been shed from the trailing edge. In order that
the circulation may be zero around the large contour we must also have
an equal and opposite circulation, $-\Gamma_v$, which will cancel the Γ_v connected with the starting vortex. This circulation, $-\Gamma_v$, is the airfoil
circulation for which we were looking. It is clear that the strength of
the " starting vortex," as it is called, will be just such as to make the
fluid flow smoothly off the trailing edge, since as long as there is flow
around the sharp edge additional starting vortices will be formed which
will cause additional circulation around the airfoil and move the rear
stagnation point farther aft along the upper surface. Visual flow
observations at Prandtl's laboratory show that the above picture is not
in the least fictitious but corresponds very closely to the conditions
actually observed.

One further remark completes the description of the flow conditions
around a finite span wing. From the Helmholtz theorem, stating that
vortices cannot end in the fluid, it is at once seen that the starting vortex
must actually be connected with the trailing vortices from the wing tips.

The final vortex picture is indicated in the sketch of Fig. 1·57.   The starting vortex of a wing moving in a stationary fluid is left behind as the wing moves forward, until finally it is so far downstream as to be effectively removed to infinity.   This starting vortex formation occurs not only when a wing is first set into motion, but also when the circulation around the wing is subsequently changed for any reason whatever.

CHAPTER 2

THE AERODYNAMICIST'S BASIC DATA

2-1. Airfoil Sections and Characteristics

Modern airfoil sections are the result of a long historical development, certain typical steps of which are indicated in Fig. 2·1. The wings for the first successful gliders were made by stretching cloth over one side of a supporting framework. Experimenters like Lilienthal

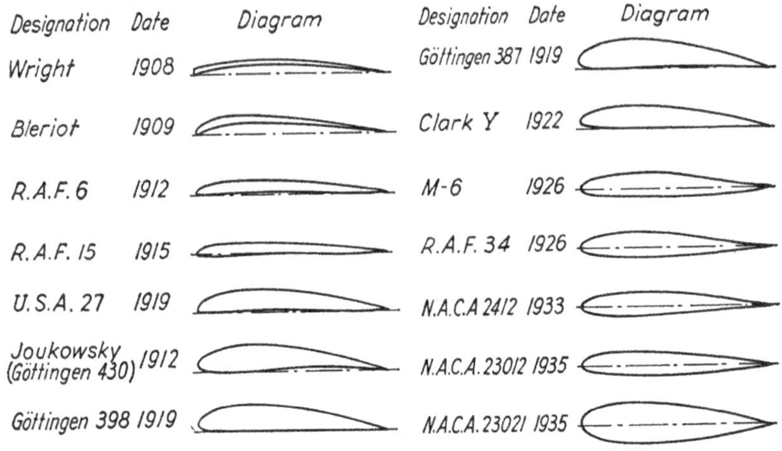

Designation	Date	Diagram	Designation	Date	Diagram
Wright	1908		Göttingen 387	1919	
Bleriot	1909		Clark Y	1922	
R.A.F. 6	1912		M-6	1926	
R.A.F. 15	1915		R.A.F. 34	1926	
U.S.A. 27	1919		N.A.C.A 2412	1933	
Joukowsky (Göttingen 430)	1912		N.A.C.A. 23012	1935	
Göttingen 398	1919		N.A.C.A. 23021	1935	

Fig. 2·1. Historical sequence of airfoil sections.

discovered very early that a certain amount of curvature or camber gave much more satisfactory results than did flat surfaces, and experience soon indicated the desirability of completely enclosing the wing framework so as to furnish airfoil profiles having a reasonably streamline shape. However for many years airfoil design was done essentially by eye and without any attempt at systemization. A very widely employed series of sections was developed in England between 1912 and 1915 (RAF 6 to RAF 15), the shapes being chosen largely for structural reasons. Modifications of these sections were worked out and generally employed in the United States following the First World War (cf. the USA 27).

The first systematic family of sections was tested at Göttingen during the First World War. This was called the Joukowsky family, from the Russian scientist who derived the shapes mathematically, using the method of conformal transformation. The airfoils were characterized by circular arc mean camber lines, large radius leading edges, and very thin trailing edges, but modifications of them developed at Göttingen were much used for many years (e.g., G 387 and G 398). In 1926 Dr. Max Munk developed and tested a family of 27 sections which were quite popular for some time. Their special feature was an S-shaped mean camber line giving a reflexed trailing edge and a stationary center of pressure (e.g., M-6). Since about 1935 systematic families of airfoils developed by the N.A.C.A. have been almost universally used in this country.

Some years ago it was customary to discuss airfoil characteristics in terms of variations in the upper and lower surfaces separately. This is

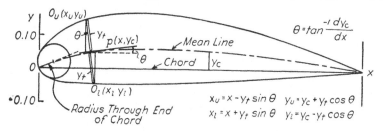

Fig. 2·2.   Notation used in airfoil section geometry.
(From *N.A.C.A. Tech. Rep.* 460.)

now recognized as a very unsatisfactory procedure since modifications to either upper or lower surface may alter the flow characteristics over the entire airfoil. The modern point of view is to discuss airfoil geometry in terms of:

(a) The mean camber line $y_c(x)$, cf. Fig. 2·2.
(b) The thickness function $y_t(x)$.

As indicated in the figure the thickness function may be considered as a fairing wrapped symmetrically around the mean camber line which serves as a skeleton of the airfoil.

It is a result of hydrodynamic theory, which is borne out by experiment, that some of the most important aerodynamic characteristics of airfoils depend almost entirely on the shape of the mean camber line and are practically independent of the thickness function, which serves essentially as a streamlining or fairing and whose optimum shape appears to be nearly independent of the camber. The thickness function affects primarily the profile drag and to some extent $C_{L\max}$. For the N.A.C.A. families under consideration a single basic thickness function is used. This has a maximum thickness of 0.20 c located 0.3 c aft of the leading edge as shown in Fig. 2·3. A family of thickness functions is obtained from this basic curve by multiplying all the ordinates by $\dfrac{t/c}{0.20}$ where

$$t/c = \frac{\text{Maximum thickness}}{\text{Chord}} \quad \text{for the thickness function desired}$$

$$0.20 = \frac{\text{Maximum thickness}}{\text{Chord}} \quad \text{for the basic thickness function}$$

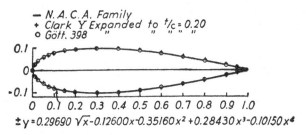

Fig. 2·3.   N.A.C.A. basic thickness function curve.
(From *N.A.C.A. Tech. Rep.* 460.)

An individual member of the family is specified by two digits giving 100 t/c. Thus 15 represents a symmetrical airfoil of the family having a maximum thickness equal to 15 per cent chord.

The first series of N.A.C.A. airfoils to be discussed * has the mean camber line of its members formed from two parabolas joined smoothly

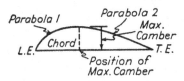

Fig. 2·4.   Mean camber function of N.A.C.A. four-digit airfoil family.

at the maximum camber point as shown in Fig. 2·4. Each mean camber line of this series of airfoils is completely defined by two numbers, one giving the magnitude of the maximum camber (in per cent of chord) and the other giving the location of this maximum camber (in tenths of chord)

---

* Eastman N. Jacobs, Kenneth E. Ward and Robert M. Pinkerton, "The Characteristics of 78 Related Airfoil Sections from Tests in the Variable-Density Wind Tunnel," *N.A.C.A. Tech. Rep.* 460 (1939).

aft of the leading edge. Accordingly each complete airfoil in the series is specified by four digits as shown below. The symmetrical members of the family are given zeros for the first two classifying digits. The 78 airfoils of the family, whose characteristics are given in *N.A.C.A. Technical Report* 460, are shown in Fig. 2·5.

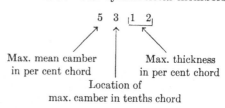

Max. mean camber in per cent chord

Location of max. camber in tenths chord

Max. thickness in per cent chord

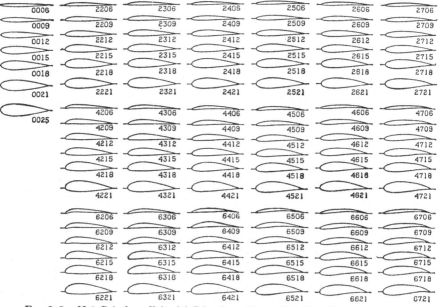

Fɪɢ. 2·5.   N.A.C.A. four-digit airfoil family.   (From *N.A.C.A. Tech. Rep.* 460.)

The second N.A.C.A. series has two subdivisions. For both the forward part of the mean camber line is a cubic. For one of the subdivisions the rear portion is a straight line while for the second it is a cubic which is concave upwards. The mean camber line of the first is called "simple," and the second is given the adjective "reflexed" because of its S-shape. Figure 2·6 illustrates these characteristics and defines the 5-digit notation used in connection with this family. The variable density tunnel results for the 5-digit series are given in *N.A.C.A. Technical Report* 610 * and the shape of the members tested is indicated in Fig. 2·7.

* Eastman N. Jacobs, Robert M. Pinkerton, and Harry Greenberg, "Tests of Related Forward-Camber Airfoils in the Variable-Density Wind Tunnel," *N.A.C.A. Tech. Rep.* 610 (1939).

The aerodynamic characteristics of airfoil sections which are usually presented and upon which the designer's choice of a section is based are:

α_{L0} = angle of attack of chord line for zero lift,

$\dfrac{dC_L}{d\alpha_0} = a_0 = 2\pi\eta$ = slope of lift curve for infinite Æ.

a.c. = aerodynamic center = point about which C_M = constant below the stall.

$C_{D0_{min}}$ = minimum profile drag coefficient.

$C_{L_{opt}} = C_L$ at which $C_{D0} = C_{D0_{min}}$.

$C_{L_{max}}$ = maximum lift coefficient.

C_{M0} = pitching moment coefficient at zero lift.

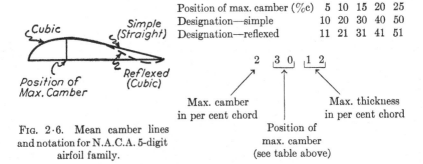

Position of max. camber (%c)	5	10	15	20	25
Designation—simple	10	20	30	40	50
Designation—reflexed	11	21	31	41	51

FIG. 2·6. Mean camber lines and notation for N.A.C.A. 5-digit airfoil family.

Most of these parameters are either familiar or self-explanatory. C_{M0} and a.c., however, require a little discussion. Hydrodynamic theory leads to the result that there should be one location of the moment axis about which C_M is constant independent of C_L, and the theory locates this so-called aerodynamic center near the 25 per cent point on the chord line. In practice the a.c. usually lies on or close to the chord line between 0.22 c and 0.25 c aft of the leading edge. The moment coefficient at zero lift, C_{M0}, is of great importance for very high speed airplanes and especially for planes designed for diving. Consider a plane in the latter category which is in a vertical dive under conditions of zero lift, so that the wing moment is determined by C_{M0}. If C_{M0} is negative, as it usually is, then the air forces produce a diving moment on the wing. This may be very large because of the large dynamic pressure involved in this kind of flight. If the plane is to be in equilibrium so as to continue its dive this diving moment must be balanced by a stalling moment arising from a download on the tail. The required download may easily be enormous if C_{M0} has a large negative value and if the speed is high. If C_{M_0} were zero, no such tail load would be required. Hence for all airplanes designed to fly at high speed and small C_L it is very important to make C_{M_0} numerically as small as pos-

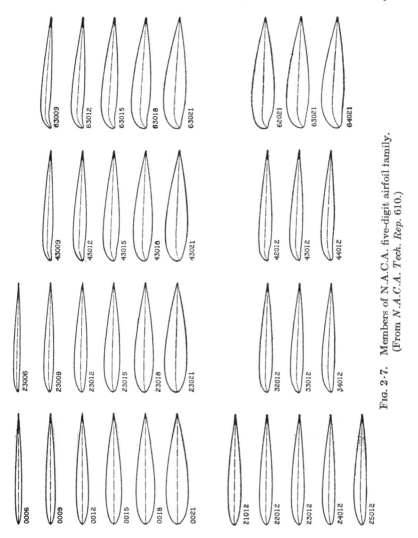

FIG. 2·7.   Members of N.A.C.A. five-digit airfoil family. (From *N.A.C.A. Tech. Rep.* 610.)

sible, with the value zero as the final objective.   Of the other parameters listed above it can only be said in general that it is desirable to have $C_{L\max}$ as large as possible and $C_{D_{0\min}}$ as small as possible.

The experimental results on the above mentioned families of airfoils lead to the following broad conclusions, several of which are also consequences of the hydrodynamic perfect fluid theory:

(a) α_{L_0} is zero for symmetrical wings and increases negatively as the camber increases. It is hardly affected by thickness.

(b) a_0 decreases slightly as the thickness increases above about 12 per cent, but for normal airfoils generally corresponds quite closely to $\eta = 0.9$. The lift curve slope is practically unaffected by camber.

(c) C_{M_0} is zero for symmetrical airfoils and increases negatively with increasing camber. Reflex or S-shape in the camber line makes C_{M_0} (when negative) numerically smaller. Considerable reflex as in the Munk sections can result in a zero value of C_{M_0}. Moving the maximum camber location forward has the same effect as S-shape, although its influence is not so powerful as that of reflexing the trailing edge portion.

(d) The aerodynamic center is close to the $0.24\,c$ point for thicknesses up to about 15 per cent. For larger thicknesses it tends to move forward slightly. It is little affected by camber.

(e) $C_{D_{0min}}$ increases fairly rapidly with thickness and more slowly with camber.

(f) $C_{L_{opt}}$ increases with camber from the value of zero which it has for symmetrical reasons.

(g) $C_{L_{max}}$ increases rapidly with camber and also with thickness up to about 12 per cent. As the thickness increases beyond this point $C_{L_{max}}$ at first remains constant and then decreases more and more rapidly.

Practically all of the above relations are exhibited graphically in *N.A.C.A. Technical Report* 460.

Within the past year or two the five-digit family has become increasingly popular, the 23000 series having been widely used. In *N.A.C.A. Technical Report* 669 * many of the earlier N.A.C.A. airfoil data have been reanalyzed and definitive values given for the major aerodynamic characteristics of most commonly used sections. In this and other recent N.A.C.A. publications a system of notation is used for the dimensionless coefficients which we have not yet discussed. In order to distinguish clearly between average coefficients for a complete wing and true

---

* Cf. Eastman N. Jacobs and Ira H. Abbott, "Airfoil Section Data Obtained in the N.A.C.A. Variable-Density Tunnel as Affected by Support Interference and Other Corrections," *N.A.C.A. Tech. Rep.* 669 (1939).

airfoil section coefficients corresponding to a two-dimensional flow over a particular profile the following two sets of coefficients are introduced:

C_L, C_D, C_M average coefficients for a complete wing.

C_l, C_d, C_m true airfoil section coefficients.

This convention has been widely adopted and often proves convenient.

Although definite rules guiding the choice of airfoil section for a particular airplane cannot be given it may be helpful to mention certain of the items which should be taken into account:

(a) The section must have a shape suitable for enclosing the necessary structural members. In general increasing the thickness decreases the weight of the wing structure.

(b) C_{M_0} should have a value as near zero as possible.

(c) $C_{D_{0min}}$ should be as small as possible and should occur at the C_L at which the plane is designed to have its best performance.

(d) $C_{L_{max}}$ should be as large as possible.

(e) The curve of C_{D_0} vs. C_L should be as flat as possible so that the profile drag will be small over a large range of C_L's.

The above and other airfoil selection criteria will be discussed more fully in the later sections on performance and stability.

2–2. Monoplane Wing Theory; Spanwise Lift Distribution

For problems connected with airplane performance and stability the results of the simple Prandtl wing theory assuming elliptic lift distribution are usually adequate. However, for stress analysis purposes it is necessary to have a fairly accurate knowledge of the spanwise distribution of lift along a wing in order that wing bending moments may be determined. Most of the modern procedures for carrying out such calculations for arbitrarily shaped wings are based on an analysis of the problem given in 1931 by Miss Lotz, one of Prandtl's assistants. The Lotz method involves very complicated and tedious Fourier series calculations, the computations for a single wing requiring some two days for an experienced engineer familiar with the procedure. Several modifications and simplifications of the Lotz method have been proposed, but all require lengthy calculations. ANC-1* gives a routine form which may be followed blindly by an engineer and which requires no understanding of the meaning of the steps taken. In this section some idea

* "Spanwise Air-Load Distribution," Army-Navy-Commerce Committee on Aircraft Requirements, April, 1938.

of the nature of the results of such calculations will be given without any attempt to follow or explain the details.

If downwash did not exist so that geometrical and effective angle of attack were equal, then the local lift coefficient $C_l = L'/qc$ would be proportional to the local geometrical angle of attack measured from zero lift, i.e.,

$$C_l = a_0(\alpha - \alpha_{L_0}) \quad \text{(neglecting downwash)} \qquad [2 \cdot 1]$$

where α and α_{L_0} refer to a particular section along the span. This is just equation $1 \cdot 34$ in which α_0 is put equal to α because downwash is neglected, and the notation C_l has been used to indicate that airfoil section rather than over-all wing characteristics are under consideration. From this equation it follows that the lift per unit length along the span is given by

$$L' = a_0 q c(\alpha - \alpha_{L_0}) \quad \text{(neglecting downwash)}$$

and since $a_0 q$ is constant over the span it follows that

$$L' \sim c(\alpha - \alpha_{L_0}) \quad \text{(neglecting downwash)} \qquad [2 \cdot 2]$$

It is customary to call a plot of L' vs. y a load distribution diagram; such a diagram is what is required for stress analysis purposes.

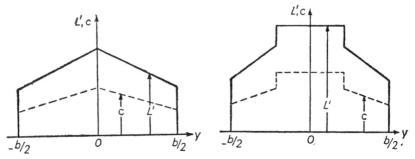

FIG. $2 \cdot 8$. Chord and load distributions for untwisted wings, neglecting downwash.

Let us consider a few typical load distribution diagrams neglecting the effect of downwash so that equation $2 \cdot 2$ is valid. If a wing has no geometrical twist then α is constant over the span and if the airfoil sections are all the same then α_{L_0} is constant across the span. From equation $2 \cdot 2$ we see that it is not α or α_{L_0} individually which are significant as far as load distribution is concerned, but rather the combination $(\alpha - \alpha_{L_0})$. Whenever $\alpha - \alpha_{L_0}$ is constant across the span, which means that all the zero lift lines are parallel, the wing is said to have zero aerodynamic twist, and in such cases L' is proportional to c, neglecting downwash. A load distribution diagram would then have exactly the same shape as a diagram giving the spanwise distribution of chord, i.e., a plan-form diagram.   In particular if the chord has discontinuities the load distribution will have corresponding discontinuities. Figure 2·8 exhibits these features.

Consider now a wing with constant chord but having a variation in $(\alpha - \alpha_{L_0})$ over the span. Exactly the same proportionality will exist between L' and $(\alpha - \alpha_{L_0})$ as was found above between L' and c. $(\alpha - \alpha_{L_0})$ may vary as a result of variation in either α or α_{L_0}. The first is obtained by geometrically twisting the wing.   The second may

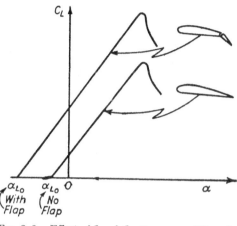

Fig. 2·9.   Effect of flap deflection on zero lift angle.

follow from a change in camber in the airfoil sections employed along the span since we have seen in section 2–1 that α_{L_0} increases negatively as camber increases. α_{L_0} is also affected by flap or aileron deflection.   Figure 2·9 gives typical C_L vs. α curves for a normal airfoil without and with a deflected trailing edge flap.   The effect of deflecting

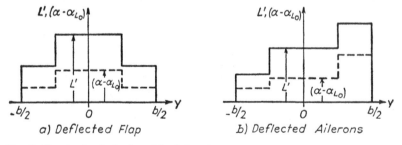

Fig. 2·10.   Angle of attack and load distributions for rectangular wings, neglecting downwash.

such a flap is essentially the same as that of increasing the camber, i.e., α_{L_0} is given an increase negatively.   Hence a wing with partial span flaps deflected or with deflected ailerons (which are essentially flaps) will have discontinuities in the curve of $(\alpha - \alpha_{L_0})$ vs. y.   Such cases are illustrated in Fig. 2·10 where the associated load distributions, neglecting downwash, are also indicated.

Now that ideas have been clarified by discussing the idealized problem in which downwash is neglected, the actual problem in which downwash occurs may be considered. Equations 2·1 and 2·2 must be modified by replacing the geometrical angle α by the effective angle of attack $\alpha_0 = \alpha - w/V$. For an untwisted wing of elliptical plan form the downwash is constant over the span so that C_l is proportional to c and the load distribution curve has an elliptical shape. For a rectangular untwisted wing the downwash is larger near the tips than at

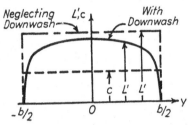

FIG. 2·11. Chord and load distributions for untwisted rectangular wing, neglecting and including downwash.

the center so that $\alpha_0 - \alpha_{L_0}$ is smaller near the tips than near the center. Accordingly the load distribution curve falls off at the tips as indicated in Fig. 2·11. This illustrates a general effect which is that the down-

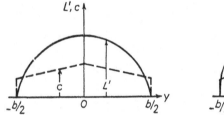
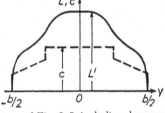

FIG. 2·12. Chord and load distributions for wings of Fig. 2·8, including downwash.

wash always adjusts itself so as to smooth out and eliminate any discontinuities in the load distribution curve. Whenever the curve of $c(\alpha - \alpha_{L_0})$ differs from an ellipse (or semi-ellipse) the downwash has the

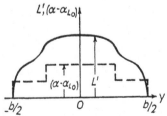
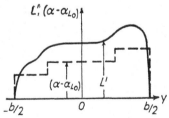

FIG. 2·13. Angle of attack and load distributions for wings of Fig. 2·10, including downwash.

effect of making the load distribution curve approach an elliptical shape, i.e., it lies between the $c(\alpha - \alpha_{L_0})$ curve and an ellipse. Figures 2·12 and 2·13 show schematically the shapes of the actual load distributions which occur when downwash is taken into account with the wings of Figs. 2·8 and 2·10, respectively.*

2-3. Multiplane Effects; Method of Images; Ground Effect; Wind Tunnel Wall Interference

It is a curious paradox of the theory of airplane wings that the calculation of the over-all characteristics of biplanes is much simpler than the determination of the distribution of load over a single monoplane wing. In fact the theory of biplanes was worked out by Prandtl's colleague Betz some years before the first satisfactory monoplane theory was developed. The reason for this apparent paradox is, however, quite plain. In dealing with biplanes the problem relates to the effects of two wings some distance apart upon one another. It is therefore possible to represent the wings by a simple, idealized arrangement of vortices. For the mono-

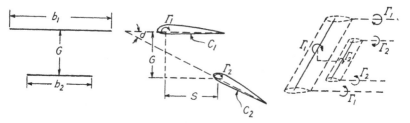

FIG. 2·14.   Biplane theory notation.

plane, however (section 1–6), such rough approximations lead to trouble and must be replaced by more complex arrangements.

In the present discussion of multiplane effects we shall, therefore, start with the problem of the biplane and shall replace each wing by a simple horseshoe vortex system as shown in Fig. 2·14. The standard conventions for gap, G, stagger, s, and decalage, d, are indicated on the sketches. The subscript $()_1$ is usually assigned to the wing of longer span and $()_2$ to that of shorter span. The lifting lines are located on the two-chord lines $\frac{1}{3}$ of the distance aft from the leading edges. Gap and stagger are defined perpendicular and parallel to the upper chord line.

Of primary importance is the induced drag which may be determined by summing or integrating the drags of each airfoil section over all of the

* Detailed discussions of this problem together with complete descriptions of calculation procedures for the determination of wing characteristics are given for tapered wings in Raymond F. Anderson's "Determination of the Characteristics of Tapered Wings," *N.A.C.A. Tech. Rep.* 572 (1936), and for similar wings with partial span flaps in H. A. Pearson's "Span Load Distribution for Tapered Wings with Partial Span Flaps," *N.A.C.A. Tech. Rep.* 585 (1937).

sections making up the airplane. For any arbitrary airfoil section the conditions are exactly those shown in Fig. 1·38. The only element of this figure which must now be interpreted differently is the downwash. Before, in discussing a monoplane wing, w could be considered as arising from the tip vortices of the monoplane. For a biplane (Fig. 2·15), at

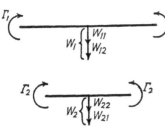

wing $(\ )_1$ the downwash, w_1, is made up of two parts: w_{11} associated with the tip vortices of wing $(\ )_1$, and w_{12} associated with the vortices of wing $(\ )_2$. Similarly at wing $(\ )_2$ the downwash w_2 is composed of a term w_{22}, associated with Γ_2, and a term w_{21} arising from Γ_1. The quantities w_{11} and w_{22} are said to be connected with self-induction, w_{12} and w_{21} with mutual-induction. From the diagram

FIG. 2·15. Downwash components for a biplane.

of Fig. 1·38 it appears that, for the induced drag of an element of wing $(\ )_1$,

$$\frac{D'_{i_1}}{L'_1} = \frac{w_1}{V} = \frac{w_{11}}{V} + \frac{w_{12}}{V} = \frac{D'_{i_{11}}}{L'_1} + \frac{D'_{i_{12}}}{L'_1}$$

The terminology of self- and mutual induction, which is also applied to these two terms, is carried over from the field of electromagnetism. By summing or integrating over wing $(\ )_1$ we would obtain

$$D_{i_1} = D_{i_{11}} + D_{i_{12}}$$

where the first term is the drag of wing $(\ )_1$ which would exist if the wing were flying by itself as a monoplane. The second term represents an additional drag due to the presence of wing $(\ )_2$ in the neighborhood of $(\ )_1$. It is therefore apparent that each wing of a normal biplane is operating less efficiently than it would if it were supporting the same load with the other wing removed. It is also clear that the greater the gap of the biplane the smaller will be the mutual induction and hence the smaller will be the unfavorable interference effects of the two wings on one another.

The calculation of the total downwash distribution over the wing span of both wings of a biplane and the integration of the resultant induced drag contributions were carried out by Prandtl's group at Göttingen. The results as presented by Prandtl are cumbersome and difficult to use, but an alternative representation developed by Munk gives the essential data in an extremely convenient form for the practical engineer.

Munk finds that it is possible to express the induced drag coefficient of any multiplane system in the form:

$$C_{D_i} = \frac{C_L^2}{\pi R_{eq}} \qquad\qquad [2\cdot3]$$

where

C_{D_i} and C_L are defined in the usual way in terms of the total forces on the multiplane and the total wing area.

$R_{eq} = \dfrac{(kb_1)^2}{S}$ = equivalent monoplane aspect ratio.

b_1 = span of the longest wing, i.e., the over-all span of the multiplane.

k = Munk's span factor

kb_1 = equivalent monoplane span.

The convenience of equation $2\cdot3$ is very apparent since it permits the calculation of induced drag for any multiplane by the determination of an equivalent monoplane wing with elliptic lift distribution which would have just the C_{D_i} vs. C_L relation of the actual airplane. For biplanes Munk's span factor k is a function of

$$\frac{G}{b_1} = \text{ratio of gap to longer span}$$

$$\mu = \frac{b_2}{b_1} = \text{ratio of shorter to longer span}$$

$$\frac{L_1}{L} = \frac{\text{lift on wing with longer span}}{\text{total lift of the biplane}}$$

Charts giving k as function of these variables are presented in several standard treatises on aerodynamics.* If the two wings of the biplane have equal chords or equal aspect ratios, k becomes a function of only two instead of three variables so that the required interpolations are much simplified.

For biplanes without decalage it is sufficiently accurate for the determination of k to assume that the ratio of lifts is the ratio of areas, i.e.,

$$\frac{L_1}{L} = \frac{S_1}{S_1 + S_2} = \frac{S_1}{S}$$

* W. S. Diehl, "Engineering Aerodynamics," The Ronald Press Co. (1936).
  K. D. Wood, "Technical Aerodynamics," McGraw-Hill Book Co. (1935).

For stress analysis purposes it is often necessary to determine the ratio of lifts more accurately. Diehl has worked out a method for doing this using a number of empirical charts. This method is discussed in detail in the works just cited.

It should be pointed out that the relation between induced drag and lift in pounds can be deduced from equation 2·3 exactly as was done for the monoplane wing. The result is

$$D_i = \frac{L^2}{\pi q (k b_1)^2} \qquad [2·4]$$

i.e., the same expression as for a monoplane except that the span is replaced by the equivalent monoplane span. In order to get an idea of the relative efficiency of monoplanes and biplanes consider a biplane made up of two equal rectangular wings. Then, as the gap varies from zero to infinity, k increases from 1 to $\sqrt{2}$. Accordingly, if b is the biplane span,

$$\frac{L^2}{\pi q (\sqrt{2}b)^2} < D_i < \frac{L^2}{\pi q b^2}$$

Consider now two rectangular monoplanes with the same wing area as the biplane and furnishing the same lift. The first has the same aspect ratio as the individual biplane wings so that its span is $\sqrt{2}b$. The second has the same span as the biplane and hence has twice the chord of the individual biplane wings. The inequalities above state that the biplane is less efficient than the monoplane having the biplane wing aspect ratio, but more efficient than the monoplane whose span equals the overall biplane span. These limits for biplane efficiency are approximately valid in other cases where the biplane does not have two equal wings. For triplanes and other unorthodox wing cellules analogous calculations have been made, the results being quite similar to those for biplanes.

There are many other types of problems in which mutual induction effects are of great practical importance. A few of them will be indicated and briefly discussed in the remainder of this section. Probably the most important is connected with the effect of horizontal tail surfaces in furnishing stability to conventional airplanes. Such surfaces are essential in order that most airplanes be capable of flight, and the stabilizing effect which these surfaces have is profoundly affected by the fact that they are operating in the downwash produced by the main wing trailing vortices. An important part of the later stability discussion will be concerned with the analysis of this problem.

A second phenomenon of less practical significance to the designer, but of some importance both to the military pilot and to the migratory bird, is the increase in efficiency which is obtained when two or more wings are in flight side by side. The essential elements of the situation are indicated in Fig. 2·16 which shows a front view of two wings flying close together with their spans lying along a single line. The notation is the same as that introduced in the biplane problem. From the diagram it is clear that an upwash is induced by each wing at the other, so that the resultant downwash at each is less than would exist if the two planes were so far apart that their interaction was negligible. In normal formation flying both of airplanes and birds this effect may cause a very appreciable decrease in the total induced drag.

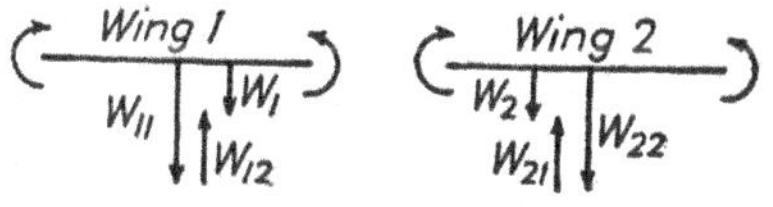

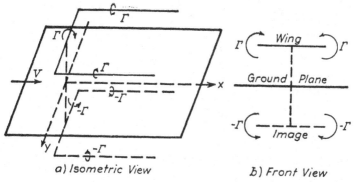

FIG. 2·16.   Downwash components for two wings side by side.

A somewhat different class of phenomena having fundamentally the same physical mechanism occurs when a lifting wing exists in the proximity of a boundary.   The simplest example is afforded by a wing flying

a) Isometric View   b) Front View

FIG. 2·17.   Horseshoe and image vortices for a wing near the ground.

close to the ground.   In order to treat this problem mathematically a device used in optics, electrostatics, and other branches of physics, is employed. This is the so-called method of images. The condition which must be satisfied at the ground plane, assuming a perfect fluid, is simply that there must be no normal component of velocity at this plane. If the wing itself is replaced by a simple horseshoe vortex then, as may be seen from Fig. 2·17, the condition of no normal component of velocity at the ground plane is satisfied if an image of the horseshoe vortex is located below the ground and has a circulation equal and opposite to that of the real wing. Conditions at the wing will now be identical for the two following cases:

(a) With the ground plane present.
(b) With the ground plane removed but the image vortex system introduced.

In view of this the interference effect of the ground on the wing may be estimated by removing the ground, introducing the image vortex, and calculating the effect of the image system on the wing. The configuration here is essentially that of the equal wing biplane except that the circulation, and hence the lift, on the lower or image wing have the opposite instead of the same sign to those of the upper or real wing. The downwash distribution is indicated in Fig. 2·18, and it appears that the presence of the ground causes an upwash at the wing. Hence when a wing is in close proximity to the ground at a given geometrical angle of attack its effective angle of attack and lift are higher, and its

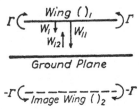

FIG. 2·18. Downwash components for a wing near the ground.

induced drag is lower, than when it is far from the ground. This abnormally low induced drag accounts for the so-called "floating" which many low-wing monoplanes experience when they approach the ground in landing. Furthermore, at least one fatal crash of a heavily loaded plane after take off has been attributed to this phenomenon. The plane in question had just enough power to take off because of the small induced drag near the ground. It did not, however, have

sufficient power to overcome the normal induced drag and hence could not climb up to any appreciable altitude where the ground effect disappeared. After skimming the surface for some time it "ran out of field," struck an obstacle, and crashed.*

Figure 2·19 shows the pressure distribution over the ground beneath a flying airplane as computed by Prandtl who used the horseshoe image picture as indicated to determine velocities, from which in turn pressures are calculated by means of Bernoulli's theorem. If the total force on the ground is computed by integrating the pressures out to infinity it is found that the total force is just equal to the weight of the airplane.

When wing models are tested in wind tunnels they are in close proximity to solid or other boundaries, and the method of images has been

* For a discussion of this accident see Elliott G. Reid and Thomas Carroll, "A Warning Concerning the Take-Off with Heavy Load," *N.A.C.A. Tech. Note* 258 (1927).

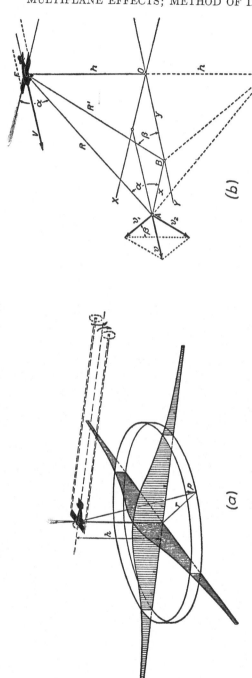

FIG. 2·19. Pressure distribution over the ground beneath a wing, and associated image system with induced velocities. (From *N.A.C.A. Tech. Rep.* 116.)

applied to this case by many investigators. All conventional wind tunnel data determined in an up-to-date tunnel are corrected for this "tunnel-wall interference" using the formulas developed from such image considerations. The final results as presented are then given as "free air" characteristics which would be expected if the tunnel boundaries did not exist. Experiment verifies these theoretical tunnel-wall corrections with a surprising degree of exactitude.

2-4. Flaps, Slots, Spoilers, Brakes, Cutouts, and Wing Mutilations

One of the most definite and serious limitations involved in conventional airfoil characteristics is the value of $C_{L\text{max}}$. This is closely related to airplane stalling speed as appears in the following equilibrium equation for horizontal flight where $L = W$ = weight (cf. equations 1·7)

$$V = \sqrt{\frac{2}{\rho} \frac{W}{S}} \frac{1}{\sqrt{C_L}} \qquad [2\cdot5]$$

The minimum or stalling speed, V_s, occurs at $C_{L\text{max}}$ and for a given weight and altitude is proportional to $1/\sqrt{SC_{L\text{max}}}$. In order to keep the stalling speed reasonably low it is necessary to make $SC_{L\text{max}}$ as large as possible. If S is made large the drag and, as a consequence, the high speed suffer, so that increasing wing area is not an entirely satisfactory method of reducing stalling speed. Variable area wings have been tried but with a few exceptions to be discussed later have never proven successful. The difficulties involved in increasing wing area to achieve slow stalling speeds emphasize the importance of large $C_{L\text{max}}$ for this purpose. It has already been pointed out that $C_{L\text{max}}$ increases rapidly with camber, but unfortunately in order to obtain really large values of $C_{L\text{max}}$ (e.g., up to 2.0) it is necessary to employ airfoil sections of such extreme camber that the resulting C_{M_0} and $C_{D_{0\text{min}}}$ values render the sections unusable for high speed flight at small C_L's. Variable camber wings have often been suggested and several have been built and flown. They in general involve a flexible structure which permits the wing to be given a high camber for slow speed flight and a small camber for high speeds. However, the danger of vibration or flutter because of the flexibility rules out the use of such wings on modern, high speed aircraft.

The trailing edge flap in any of its usual forms is essentially a device for changing the effective camber of a wing, as has already been suggested in section 2-2. Figure 2·20 shows typical three-component characteristics of a wing without and with flap. It will be noticed that with deflected flap $C_{L\text{max}}$ is much increased as is $C_{D\text{min}}$, α_{L_0} is reduced while α_{stall} is only slightly affected, and C_{M_0} is greatly increased negatively.   All these effects are similar to those of increasing camber.   The increase in C_D is larger than that which would correspond to a camber increase sufficient to produce the given $C_{L\mathrm{max}}$ increment, as would be expected from the poor "streamline" nature of the wing with deflected flap.   This large C_D increment associated with flaps is nearly as useful for some purposes as the $C_{L\mathrm{max}}$ increment.   It steepens the gliding angle as will be shown in the section on *Performance*, and also causes a landing plane to decelerate more rapidly and so greatly reduces the "floating" tendency of very clean airplanes.

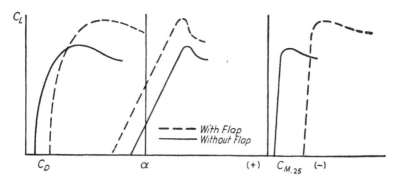

Fig. 2·20.   Typical aerodynamic characteristics for a wing without and with flap.

Figure 2·21 gives the geometry and major aerodynamic characteristics for a number of types of flap in common use.   The plain and slotted flaps are often used where low drag at high lifts is required as in take-off.   The split flap is simple and gives large increments in $C_{L\mathrm{max}}$ and in C_D.   The Fowler flap gives the highest lifts, which may be partially explained by the increase in wing area which accompanies its deflection.   Its worst feature is the extremely large diving moment which accompanies its use.   In general it may be said that the modern high speed airplane of very high wing loading would be completely impractical without flaps.   Accordingly some form of flap is fitted to practically every airplane which answers this description.

A high lift device of an entirely different type is the slot, which was really made practicable by Lachmann and Handley-Page.   Such a slot in its open and closed positions is shown schematically in Fig. 2·22.   In the closed position the slot forms the wing leading edge and gives the airfoil a quite normal section.   When open, near the leading edge a slot is formed through which air passes at high speed from the pressure to the suction side of the wing.   As indicated in Fig. 2·23 the slot has no appreciable effect on the wing lift below the normal stall, but it does lead to a higher $C_{L_{max}}$ by postponing the burble point to a considerably higher angle of attack. The mechanism by which this delaying of separation is accomplished is rather complex, but one of the important

Designation	Diagram	$C_{\ell max.}$	α at $C_{\ell max.}$	$-C_{M_0}$
Basic Airfoil		1.54	15.5°	0.01
With 0.2C Split Flap Deflected 60°		2.53	12°	0.18
With 0.2C Plain Flap Deflected 60°		2.38	12.5°	0.22
With 0.2C Slotted Flap Deflected 50°		2.76	13.5°	0.25
With 0.27C Fowler Flap Deflected 30°		2.90	10.5°	0.42

FIG. 2·21.   Characteristics of several flaps with N.A.C.A. 23012 Airfoil.
(From *N.A.C.A. Tech. Rep.* 664.)

elements is that the jet of high speed air leaving the slot accelerates the retarded boundary layer over the upper surface and permits it to continue downstream against an abnormally high adverse pressure gradient

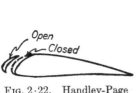

FIG. 2·22.   Handley-Page slot.

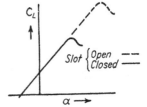

FIG. 2·23.   Lift versus angle of attack for a wing without and with slot.

before separation occurs.   The special feature of the Handley-Page slot is that it is so mounted on the wing that at low angles of attack it is held closed by the pressures over the leading edge, while at large angles of attack it is automatically sucked out into the open position.

As a high lift device to reduce landing speed the slot has not been widely used because of the difficulty of constructing landing gears permitting the attainment of the large angles of attack required for the high lifts. However, in connection with flaps it has been so employed on a number of airplanes. Its most widespread use has been on wing tips so that these portions of the wing containing the ailerons may remain unstalled after the maximum lift of the rest of the wing has been attained.

Designation	Diagram	$C_{L\,max.}$	α at $C_{L\,max.}$	Reference (N.A.C.A.)
Basic Airfoil		1.29	15°	Tech Rep 427
With Fixed Slot		1.77	24°	''
With Fixed Slot and Slotted Flap		2.26	19°	''
Basic Airfoil		1.27	14°	Tech Note 459
With Handley-Page Slot		1.84	28°	''
With Handley-Page Slot and Fowler Flap		3.37	16°	''

FIG. 2·24. Characteristics of Clark Y wing with various slots and flaps.

In this way lateral control can be maintained even above the apparent stall, which is not true with most normal unslotted wings. Figure 2·24 gives the characteristics of certain combinations of slots and flaps which have been investigated.

It is sometimes desirable to decrease lift instead of increasing it as is accomplished by flaps or slots. Two types of "spoiler" which have been used for this purpose are shown schematically in Fig. 2·25. Such devices have been chiefly employed to replace ailerons in giving lateral control, but have not yet been developed in an entirely satisfactory form for this purpose.

A special device which has been found necessary with dive bombers is some sort of air brake to increase the airplane's resistance and so decrease its diving speed. Such brakes should not greatly affect either lift or pitching moment, so as not to alter the airplane's flight path when they are operated. Two types of air brake satisfying these requirements are

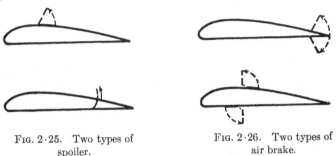

FIG. 2·25.   Two types of
spoiler.

FIG. 2·26.   Two types of
air brake.

illustrated in Fig. 2·26. Both have been used on recent production combat planes.

Some mention must be made of the effect of various types of mutilations, cutouts, and other modifications which have deleterious effects on wing characteristics. In general it is possible to alter the trailing edge region very considerably without serious sacrifice to wing aerodynamic properties. Even fairly large cutouts, which are sometimes made in biplane upper wings to improve pilots' vision (cf. Fig. 2·27), do not have

FIG. 2·27.   Wing trailing edge
cutout.

very serious aerodynamic effects if they are located in the trailing edge. However, any modification at or near the leading edge may cause premature separation which not only affects the wing behind the modification, but often spreads laterally so that a large portion of the wing is stalled, with a consequent important increase in drag and decrease in lift. In this respect gasoline tank caps, landing lights, access or inspection doors, etc., placed near the leading edge are often serious offenders.

An especially deleterious modification or mutilation of a wing is anything which causes the lift to fall to zero at some point between the two wing tips. Even a very small fore-and-aft, or longitudinal, slot makes the lift fall to zero and introduces two new trailing vortices (cf. Fig. 2·28). Not only is the lift greatly reduced, but the induced drag is enormously increased. In one case which occurred some years ago in a small training plane, sealing such a slot, which was perhaps one-half inch wide,

doubled the rate of climb.   An unfortunate junction between the wing
and fuselage of a low-wing airplane, even though no gap in the wing
exists, may have similar deleterious effects.   If the junction is good the
wing lift will carry across the
fuselage and the lift distribution
will be essentially the same
whether or not the fuselage is
mounted on the wing. If, however, the junction is bad premature separation will occur at the
wing-fuselage intersections at very
low angles of attack.   The lift
may then drop nearly to zero over
the wing portion covered by the
fuselage, additional trailing vortices appear which give a large
drag increase, and a situation much
like that of Fig. 2·28 occurs.   It

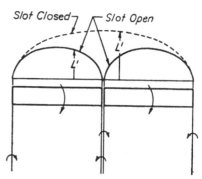

Fig. 2·28.   Trailing vortices and lift distribution for a wing with longitudinal
slot.

is often possible to transform an unsatisfactory wing-fuselage intersection into a completely satisfactory one by means of properly designed
fillets.

2–5. Parasite Drag; Scale and Compressibility Effects

Two somewhat different definitions have been used in the past for the
parasite drag of complete airplanes.   The older considers as parasite
drag that portion of the total drag of an airplane which is not associated
with lifting elements, in other words, all except wing drag.   The more
recent and in many respects the more convenient considers parasite drag
as total drag minus induced drag.   In this work we follow the second
definition which differs from the first primarily in the inclusion of wing
profile drag in the parasite drag.   The parasite drag coefficient will
accordingly be defined by the relation:

$$C_{D_p} = C_D - \frac{C_L^2}{\pi \mathcal{R}_{eq}} \qquad [2·6]$$

The actual parasite drag for an airplane is connected with this coefficient
by the usual formula

$$D_p = C_{D_p} q S \qquad [2·7]$$

where S is the wing area as customarily defined.

A considerable number of other coefficients of different form have
been used for the representation of the parasite drag of airplanes and

their components. Two will be introduced here since they will be employed later. The first is useful in describing the parasite drag contributions of various component airplane parts such as nacelles, fuselages, tail surfaces, etc. It is called the proper drag coefficient and is defined by

$$C_{D_\pi} = \frac{\Delta D}{qS_\pi} = \text{Proper drag coefficient} \qquad [2\cdot8]$$

where

ΔD = contribution of the part in question to the total parasite drag of an airplane.

S_π = proper area of the part. For nacelles, fuselage, etc., it is taken as the total projected frontal area of the part including any wing area covered by the part. For tail surfaces it is taken as the usual tail surface area not including area covered by fuselage.

The second drag coefficient to be introduced is especially convenient for performance calculations. It is not dimensionless but has the dimensions of an area and is defined by

$$f = \frac{D}{q} = \text{Equivalent parasite area} \qquad [2\cdot9]$$

It can be used either for airplane components, where D represents the drag of the individual component in question, or for a complete airplane, where D is the total parasite drag, D_p, and

$$f = C_{D_p}S \qquad [2\cdot10]$$

In turning from notation to the parasite drag itself it is recalled that drag forces, like all aerodynamic forces, may be divided into two categories which have been called skin friction and pressure drag. The first is always present in practical aerodynamics but is most conveniently discussed in connection with flow along a flat plate set parallel to the flow, where it is the only force which acts. This type of flow has already been considered in section 1-7 and has been the subject of both theoretical and experimental study by a great number of investigators. The essential features have been outlined in the earlier boundary layer discussion. For the technically important range of Reynolds numbers they include a boundary layer of retarded flow whose thickness is proportional to $\sqrt{l}$ where l represents length along the plate downstream from the leading edge (cf. equation 1·39 and Fig. 2·29). As the boundary layer becomes thicker in going downstream $(du/dy)_0$ decreases, so that the local shearing stress, $\tau_0 = \mu(du/dy)_0$, gets smaller and smaller as the distance from the leading edge increases. The total skin friction is obtained by integrating τ_0 over the entire plate, so that as l increases the average skin friction over the plate decreases. The quantitative

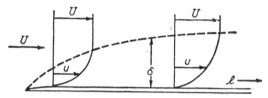

FIG. 2·29.   Laminar boundary layer along a flat plate.

expression for this was worked out in 1911 by Blasius who gave the formula for the skin friction coefficient

$$C_F = \frac{\text{Skin friction}}{(\rho/2)U^2 A} = \frac{1.328}{\sqrt{R_l}} \qquad \text{Blasius friction law} \qquad [2\cdot11]$$

where

A = total "wetted" surface exposed to the fluid

$R_l = \dfrac{Ul}{\nu}$

l = length of the plate from leading to trailing edge

This is all relatively simple and easy for the engineer to work with. Unfortunately, in practice another factor enters which has been ignored in our discussion and which seriously complicates matters. For reasons not yet entirely clear, when R_l reaches some critical value (normally between 200,000 and 2,000,000) the so-called laminar flow which has been assumed becomes unstable and an entirely new type of flow appears. This is called "turbulent flow" and has the characteristic that the fluid, instead of moving smoothly in layers or lamina which slip over one another, has an irregular, eddying, or fluctuating nature. The flow is filled with eddies which move irregularly about producing rapid fluctuations in velocity. These fluctuations have such a high frequency that ordinary Pitot tubes and other hydraulic measuring instruments do not record or respond to them at all. Special hot-wire anemometers with extremely rapid response are required in order that the fluctuating character of the flow may be made apparent. The average velocity, which is indicated by ordinary instruments, may be denoted by $\bar{u}$ and is the quantity with which the aerodynamicist is primarily concerned. Unfortunately the Newtonian friction law does not hold for this mean velocity, i.e., $\tau \neq \mu(d\bar{u}/dy)$. It is clear that this must be so since the small turbulent eddies moving from layer to layer must carry momentum across between adjacent layers and give a very large dragging effect.   In fact, if we write formally, for the case of turbulent flow,

$$\tau = \epsilon \frac{d\bar{u}}{dy}$$

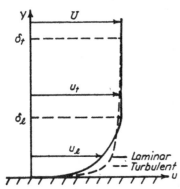

FIG. 2·30.   Laminar and turbulent velocity profiles.

the coefficient ϵ, which is analogous to the viscosity coefficient μ and which is called the exchange coefficient, is not a constant of the fluid but may have a value ranging from 30 to 1,000,000 times that of μ, depending on the Reynolds number and other parameters.   In other words the effective or apparent viscosity is very much larger in turbulent than in laminar flow.   It would be expected, therefore, that the boundary layer characteristics would be quite different in the two cases, and this is found to be true.   The most important differences are the following:

(a)   The turbulent boundary layer is much thicker than the laminar one and the thickness increases more rapidly with l.

(b)   The turbulent velocity profile is much fuller near the wall and flatter some distance from the wall (cf. Fig. 2·30).

(c)   The skin friction is much larger in turbulent than in laminar flow.

It should be pointed out that extremely close to the wall the turbulent fluctuations are damped out, so that the motion here is again laminar. This region has been described as the laminar sublayer or laminar film. Its thickness is a very small fraction of the turbulent boundary layer thickness.

Von Kármán has given a theory of the turbulent boundary layer which has been accurately verified by experiment and which leads to the formula for the skin friction of smooth, flat plates:

$$\frac{0.242}{\sqrt{C_F}} = \log_{10}(R_l C_F) \qquad \text{von Kármán logarithmic skin friction law} \qquad [2 \cdot 12]$$

Figure 2·31 gives a plot of the laminar and turbulent skin friction coefficients plotted against Reynolds number (from equations 2·11 and

2·12) and shows clearly how much lower the former is, particularly at large Reynolds numbers.

As mentioned above the laminar flow in the boundary layer along a flat plate breaks down at some critical value of R_l and the flow downstream from this point is turbulent. The point along the plate at which this occurs is called the transition point, although in general the transition occurs over a region of finite length rather than at a definite point. The transition point corresponds to a critical value of the Reynolds number, $R_{l\text{crit}}$, which value depends on the roughness of the

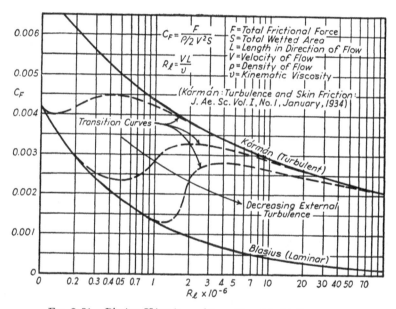

FIG. 2·31.   Blasius, Kármán, and transition skin friction curves.

plate surface, the amount of turbulence or fluctuation in the outside free stream, etc. Since $R_{l\text{crit}} = (Ul/\nu)_{\text{crit}}$ is approximately constant for a given plate and windstream it follows that the transition point in general moves towards the leading edge as the flow velocity, U, increases as shown in Fig. 2·32. This motion of the transition point with velocity means that for a given plate, as U is changed the relative proportions of the plate having laminar and turbulent boundary layers also change. The skin friction curve for such a plate therefore follows a transition curve between the Blasius and Kármán curves as R_l changes, lying on or close to the Blasius curve at small R_l's and approaching the Kármán curve at large R_l's where the transition point lies close to the leading edge so that practically all the boundary layer is turbulent. The particular transition curve which is followed depends largely on the degree of turbulence in the outside free stream, a very smooth flow giving a high $R_{l_{crit}}$ and a late transition, and vice versa. Typical transition curves are shown in Fig. 2·31.

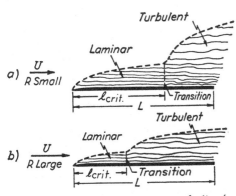

The rather complicated skin friction characteristics outlined above for a flat plate appear in a very similar form for bodies with curved surfaces. In particular for airfoils at low angles of attack the drag is practically pure skin friction and the formulas and curves given above may be

Fig. 2·32. Effect of free-stream velocity (or Reynolds number) on transition.

applied to the estimate of profile drag. In making such estimates it is usually accurate enough to neglect the variation in U over the surface and assume $C_{D_0} = 2C_F$. The factor 2 appears because C_F is defined in terms of wetted area while C_{D_0} is based on the conventional projected wing area.

In addition to furnishing quantitative data on skin friction the boundary layer picture leads to an explanation of certain very important characteristics which are connected with the pressure drag of bluff bodies. As a typical example of these phenomena consider a flow with uniform velocity past a sphere or circular cylinder, so that the diagram of Fig. 1·49 represents conditions in the boundary layer. In drawing this figure it was implicitly assumed that the flow in the boundary layer was laminar as is indicated by the shape of the velocity profiles. If the boundary layer were turbulent instead of laminar all the profiles before the separation would be much fuller, i.e., the velocity near the wall would be higher (cf. Fig. 2·30). It would be expected that in such a case a particle of fluid just outside the surface would be able to proceed further downstream against the adverse pressure gradient before being brought to rest. In other words a turbulent boundary layer should resist separation better than a laminar one, and the separation point for a turbulent layer should be farther downstream than that for a laminar layer. Experiment shows that this is in fact the case, the flow around a sphere with laminar and turbulent boundary layer, respectively, being shown schematically in the two sketches of Fig. 2·33.

From the point of view of parasite drag the essential difference between these two diagrams is that the turbulent wake is very much broader in case (a) than in (b). This means that in the latter case the flow approaches much more closely to the perfect fluid flow with high pressure on the rear face than in (a), and the suction associated with the wake acts over a much smaller surface. The pressure drag should therefore be much smaller for the flow with turbulent boundary layer than for that with laminar layer. Wind-tunnel tests show that the drag in (b) is actually less than one-fifth as large as that in (a).

The question immediately arises as to when these two types of flow, having such strikingly different characteristics, may be expected to

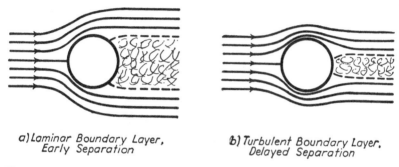

a) *Laminar Boundary Layer,*
Early Separation

b) *Turbulent Boundary Layer,*
Delayed Separation

Fig. 2·33.   Flows around a sphere having laminar and turbulent boundary layer.

occur. Here the same physical mechanism as controls transition along a flat plate is effective. For a given airstream and surface condition transition in the sphere boundary layer would occur at an approximately constant value of $R_l = U_l/\nu = (U_l/\nu)_{crit}$, where l is distance downstream along the surface from the stagnation point. Hence as U_0 (or $R = U_0 d/\nu$, where d = sphere diameter) increases, the transition from a laminar to a turbulent regime should move forward along the surface of the sphere. As long as the transition point would occur downstream of the laminar separation point the flow will have the "laminar boundary layer" character shown in Fig. 2·33 (a). However when the transition point moves forward until it is upstream of the laminar separation point, the boundary layer in this region becomes turbulent and resists separation, so that the separation point jumps downstream and the flow takes on the character shown in Fig. 2·33 (b). Accordingly as R increases from a low value the drag coefficient remains roughly constant at a relatively high value corresponding to a flow like (a). When R reaches a definite critical value, R_{crit}, the transition point coincides with the laminar separation point, the flow character changes to that of type (b), and the drag coefficient drops. As R increases still further C_D remains roughly constant at a relatively low value. Each of the curves of Fig. 2·34 corresponds to this course of events for a smooth sphere in a given windstream. The value of R_{crit} has been arbitrarily defined as corresponding to $C_D = 0.3$. The final complication in this already complicated phenomenon is the fact that R_{crit} depends very markedly on the sphere surface texture and on the degree of steadiness or turbulence in the airstream. As the surface roughness or the airstream turbulence increases, the value of R_{crit} decreases, i.e., the transition to the turbulent boundary layer flow occurs earlier. Unfortunately, all wind-tunnel airstreams have a certain amount of turbulence or unsteadiness so that considerable variations

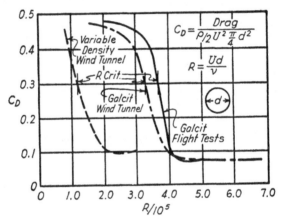

FIG. 2·34. Drag coefficient for a sphere.

are sometimes found in R_{crit} when results from different tunnels are compared.

The phenomenon of a "critical Reynolds number," at which the drag coefficient changes suddenly, is not confined to spheres or circular cylinders but occurs with many other bodies. For spheres and circular cylinders it occurs just in the range of Reynolds numbers, which is of technical importance, but for streamline shapes it fortunately appears at very low Reynolds numbers which are below those of practical importance in aeronautics. It is obvious that in making extrapolations from model to full scale conditions it is necessary to be sure that such a critical point does not lie in the region over which the extrapolation is extended. This is one of the chief factors which places a lower limit on the speed and size of wind tunnels for use in airplane design.

Most standard treatises on aerodynamics include a large amount of information on the parasite drag of various component parts of airplanes, the data having been collected from an immense number of wind-tunnel tests in various wind tunnels.   No attempt will be made to reproduce this material here, but a few proper drag coefficients for the major components of modern monoplanes are listed at the end of this section.   The values were obtained from a statistical study of the results of some 200 different wind-tunnel investigations carried out in the GALCIT wind tunnel.   They represent drag added when the element is added to the other parts of an airplane.

The problem of estimating full-scale parasite drag values for a new airplane design is one of the most difficult which the aerodynamicist has to tackle.   The following procedures have been widely used:

(a) Summation of the drags of the various components using collections of data like those contained in the standard aerodynamic reference books, or more roughly from the following table of average GALCIT values.

(b) Extrapolation of wind tunnel model test results on an accurate model to full-scale Reynolds number and surface texture conditions.

(c) Comparison of the new design with similar existing airplanes for which accurate data are available from flight test observations.

All three methods require great skill and experience if they are to lead to accurate results, and the most successful aerodynamicists usually use a combination of all three, in addition to drawing considerably on their past experience.   For very high-speed airplanes (of the order of 400 m.p.h.) another factor, in addition to those mentioned above, must be considered.   This is the effect of compressibility, which may cause serious drag increases at such speeds, particularly at corners or places where the radius of curvature is small.   The data required in this connection are unfortunately too meager as yet to permit any general discussion of the problem.   Additional data, should however, become available in the near future.

PROPER DRAG COEFFICIENT OF MAJOR AIRPLANE COMPONENTS

(GALCIT Statistical Data, R $\sim 10^6$ to 2×10^6 based on wing chord)

Element	Description	C_{D_π}
Fuselage	Dirigible hull, circular section, tested alone	0.072
Fuselage	Large transports or bombers, no nose engine or turrets	0.070 to 0.105
Fuselage	Small plane, including nose engine, closed cockpit	0.090 to 0.130
Nacelle	Nacelle mounted above a wing	0.250
Nacelle	Leading edge nacelles, small airplane, relatively large nacelles	0.120
Nacelle	Leading edge nacelles, large airplane, relatively small nacelles	0.080
Tail Surfaces	Single-engined, low-wing monoplane	0.0085 to 0.0120
Tail Surfaces	Multi-engined, low-wing monoplane	0.0060 to 0.0110
Tail Surfaces	Single-engined, high-wing monoplane or biplane	0.0120 to 0.0180

CHAPTER 3

PERFORMANCE

3–1. Introduction; Standard Atmosphere; Equilibrium Equations; Fundamental Performance Equation

In this chapter we shall be concerned primarily with a group of problems which are sometimes referred to under the title of "classical airplane performance." Included are:

(a) The maximum velocity for unaccelerated level flight, V_{max}.
(b) The velocity for maximum steady rate of climb, V_c.
(c) The value of the maximum steady rate of climb, C.

All are considered as functions of the altitude h, and correspond to arbitrarily selected values of engine power. The minimum time required to climb to an altitude h is sometimes included in the classical performance problem. Certain "special" performance problems such as range and take-off will also be discussed. Considerations of stability and control, however, do not enter.

As indicated above, the dependence of certain performance elements on altitude is important. However, the propulsive and aerodynamic characteristics which determine these elements of performance do not depend on altitude itself but rather on air density, pressure, and temperature. In order to make the problem precise, it is accordingly necessary to specify a definite relation between these factors. This has been done in the "International Standard Atmosphere" which has been almost universally adopted by the major countries of the world and which will here be briefly discussed. There are two fundamental relations which underlie atmospheric characteristics. The first is the "perfect gas" equation of state, $p/\rho = kT$, where k is the so-called gas constant and T is the absolute temperature. The second is the equilibrium equation for an element, which equates the weight of air in a volume element of height, dh, with the increase in pressure, dp, in moving down a distance $-dh$: $dp = -\rho g dh$. In view of these two relations it is only necessary to specify one additional relation between the four variables p, ρ, T, and h in order that p, ρ, and T be completely determined as functions of h. This required relation is furnished by the so-called linear lapse rate T (°F.) $= 59 - 0.003566\,h$ (ft.) which has been empirically chosen as the basis for the standard atmosphere after a study of average lapse rates observed all over the world.

The "Density and Pressure Altitude Conversion Chart" is essentially a convenient graphical representation of the above atmosphere relations. Fundamentally it is a chart expressing the equation of state, $p/\rho = kT$ as shown in Fig. 3·1. On this chart is drawn a line giving

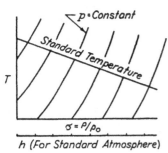

FIG. 3·1. Schematic atmosphere chart.

the temperature-pressure relation of the standard atmosphere. This permits the construction of the auxiliary altitude scale which is shown below the chart and which is based on the equilibrium equation. As the chart is actually drawn the constant pressure lines are labeled in feet and called constant "pressure altitude," h_p, lines. This is because airplane altimeters, which measure pressure, are calibrated to read in feet rather than in inches of mercury or some other unit of pressure.

The calibration is so chosen that if the altimeter were in the standard atmosphere it would read the actual altitude. In other words, h_p is really the altitude in the standard atmosphere at which a given pressure would exist. Similarly the density scale is labeled in feet of "density altitude," h_d. h_d is the height in the standard atmosphere at which the density ratio, ρ/ρ_0, has some specified value σ. Figure 3·2 represents such a chart in a form convenient for engineering calculations.* Auxiliary scales giving pressure in millimeters of mercury, density ratio σ, and $\sqrt{\sigma}$ are included, as well as curves giving the kinematic viscosity ratio ν/ν_0. Chart A is an enlarged version of this figure which has proven very useful for actual computation purposes.

Performance calculations are practically always made using the standard atmosphere so that $h_d = h_p = h =$ true altitude, and p, σ, and T are given as definite functions of h. The data for the standard atmosphere may be obtained from Fig. 3·2 or Chart A. When flight tests are made the atmospheric conditions are almost never exactly those of the standard atmosphere, so that it is necessary to correct the flight test observations to furnish values which would have been obtained if con-

* It is believed that this form of chart was first developed for aeronautical use by engineers of the Douglas Aircraft Co.

ditions had been standard. We shall later return to a further discussion of this matter.

An analysis of the performance problem itself begins with the consideration of the equilibrium conditions for the steady, rectilinear flight

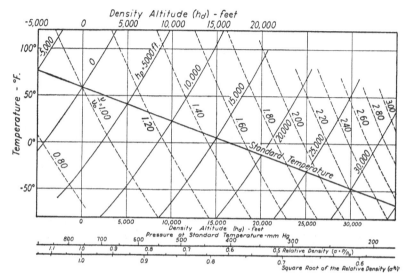

Fig. 3·2. Density and pressure altitude conversion chart.

of an airplane. The velocity of the airplane along its flight path is taken as V, the inclination of the flight path upwards from the horizontal as θ, and the controls are assumed to be set so that there is no

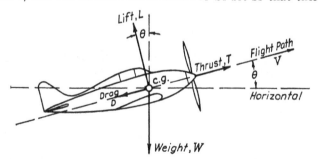

Fig. 3·3. Equilibrium diagram for steady rectilinear flight.

resultant moment of the air forces about the center of gravity. The external forces may then be resolved into the four components illustrated in Fig. 3·3. Since the motion is steady, i.e., unaccelerated, the equilibrium conditions applied to the directions parallel and perpendicular to the flight path lead to the two equations:

$$\left.\begin{array}{c} L = W \cos \theta \\ T = D + W \sin \theta \end{array}\right\} \qquad [3\cdot1]$$

Here L = lift
  D = drag
  W = weight
  T = thrust due to the propeller

One simplification which is almost universally made in these equations is that $\cos \theta$ is considered equal to 1. The maximum angle of climb for normal airplanes is small enough so that this simplification leads to no appreciable error except in steep diving flight, which is not included in what we have called the classical performance problem. The first equation then becomes

$$L = W = C_L \tfrac{1}{2}\rho V^2 S \qquad [3\cdot2]$$

which determines C_L as a function of V and σ. It will appear as we proceed that the basic independent variables of the problem are V and σ or (since we always assume the standard atmosphere so that σ is a definite function of h) V and h. Since $C_L(V, h)$ is known from equation $3\cdot2$ $C_D(V, h)$ can be obtained from a polar diagram of the airplane minus propeller. This polar may have been obtained from wind-tunnel model tests or by summation of the various estimated drag components. T is determined as a function of V and h from the engine and propeller characteristics as will be shown in section 3–4 below. All the terms in the equilibrium equations have therefore the form of known functions of V, h, and θ, so that all that remains is the working out of the consequences of these equations.

It turns out, however, as so often in physical problems, that it is more convenient to work with an energy equation than with a relation between forces such as that given by equation $3\cdot1$. Such an energy equation is obtained if the second equation of $(3\cdot1)$ is multiplied by V, giving

$$TV = DV + WV \sin \theta \qquad [3\cdot3]$$

The first term represents the useful power delivered by the propulsive system and may be written

$$TV = P\eta \qquad [3\cdot4]$$

where P = power delivered by the engine to the propeller,
  η = propulsive efficiency.

The engine power P may be considered as a given quantity, so that equation 3·4 serves to define the propulsive efficiency, η.

Since the altitude of the airplane is denoted by h, $V \sin \theta = dh/dt$ = rate of climb. Equation 3·3 may then be rewritten in the form

$$W \frac{dh}{dt} = P\eta - DV \qquad [3·5]$$

which is usually taken as the fundamental equation of airplane performance. The left side represents the rate of increase of potential energy, the first term on the right, the so-called "power available" from the propulsive system, P_a, and the last term the "power required" for level flight, P_r. The right side as a whole is often called the "excess power," and equation 3·5 then expresses the relation between rate of climb, weight, and excess power. Symbolically it may be written in the form

$$\left.\begin{array}{l} W \dfrac{dh}{dt} = P_a - P_r \\[2mm] P_a = P\eta = TV \\[2mm] P_r = DV \end{array}\right\} \qquad [3·6]$$

Consistent units must, of course, be used. If the left side is expressed in foot-pounds per second and the right side in horsepower, the conversion factor, 1 hp. = 550 ft.-lb./sec., must be introduced on one side or the other.

From the discussion following equation 3·2 it is obvious that $P_a = TV$ may be expressed as a definite function of V and σ using the engine and propeller characteristics which are assumed to be given. Furthermore $P_r = DV$ can also be expressed as a function of V and σ from the polar, C_D vs. C_L. Equation 3·6 is therefore entirely analogous to the original equilibrium equation, and is actually a much more convenient relation for the purpose of performance analysis.

3–2. Simple Graphical Analysis of the Performance Equation

The procedure outlined in this section was for many years almost universally employed in routine performance calculations. Although it has now been replaced to a considerable extent by more powerful or rapid methods, it must still be returned to in certain troublesome and abnormal cases. Furthermore, the visual picture of simple performance characteristics which it furnishes is of fundamental importance, and should always be present in the mind of the worker in this field.

In this analysis the drag and power available characteristics of a given airplane are assumed to be known as functions of the velocity V and the "standard altitude" h. The classical performance problem may now be stated as follows: Given the power available and power required as functions of the independent variables V and h (or σ), determine the velocities for level flight and for maximum rate of climb, and the values of this rate of climb as functions of the same variables. The graphical procedure under discussion begins by considering a fixed value of h, say h_1, and plotting power available and power required as functions of velocity. This furnishes a diagram like that shown in Fig. 3·4.

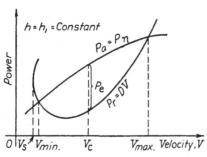

FIG. 3·4. Power-available and -required diagram.

Consider first, level flight, $dh/dt = 0$; it is clear from equation 3·6 that this will occur where $P_a = P_r$, i.e., at the intersections of the two curves. The maximum speed for level flight, V_{max}, corresponds to the intersection to the right, while the corresponding minimum speed, V_{min}, is given by the intersection to the left. The stalling speed, V_s, shown in Fig. 3·4 as smaller than V_{min}, is the lowest speed at which the airplane could fly if the power available were allowed arbitrarily large values. The illustration, in which V_s is less than the minimum level speed, is representative of all airplanes if the altitude is great enough. However, for normal, modern planes at moderate altitudes the power available is sufficiently great so that V_{min} and V_s coincide.

Equation 3·6 indicates that the conditions for maximum rate of climb will be satisfied when the excess horsepower has the largest possible value, i.e., when the vertical distance between the two power curves is a maximum, or their slopes are the same. The velocity at which this occurs is called the "speed for best climb" and is indicated by V_c on the figure. The corresponding maximum excess power is indicated by P_e. From equation 3·6 it follows that the maximum rate of climb is given by

$$C = \left(\frac{dh}{dt}\right)_{max} = \frac{P_e}{W} \qquad [3\cdot7]$$

where rate of climb, power, and weight must, of course, be expressed in consistent units. If the units are not consistent numerical constants of proportionality must be included in equation 3·7.

The procedure just outlined furnishes the maximum and minimum speeds for level flight, the speed for best climb, and the maximum rate of climb for the chosen altitude h_1. In order to obtain these quantities at other altitudes, the procedure may be repeated for other values of h. In general, as in the typical example illustrated in Fig. 3·5, the power-required curve moves up and to the right as the altitude increases, while above some critical altitude the power-available curve moves down. It follows that above the critical altitude the speed range, $V_{\max} - V_{\min}$, and the maximum excess power, P_e, decrease with increasing altitude. Finally an altitude is reached at which the power available and required

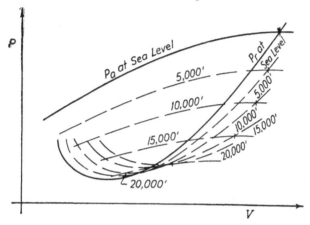

FIG. 3·5.   Power curves for the classical performance problem.

curves are just tangent; i.e., $V_{\max} = V_{\min}$, and $P_e = 0$. This altitude, at which the maximum rate of climb is zero, is called the absolute ceiling, H, if the power available for which the calculations are made corresponds to the maximum possible output of the engines.

We now have all the elements of the classical performance problem except the minimum time, T, to climb from sea level to an arbitrary altitude, h. This is obviously given by

$$T = \int_0^T dt = \int_0^h \frac{dh}{(dh/dt)_{\max}} = \int_0^h \frac{dh}{C} \qquad [3·8]$$

which can be evaluated numerically or graphically when C is known as a function of the altitude h. The curve of C vs. h obtained from the graphical analysis can sometimes be satisfactorily approximated by a straight line or a parabola, in which case the integral of equation 3·8 can be evaluated analytically.

The final results of the complete analysis may be collected in the form of a classical performance diagram such as that shown in Fig. 3·6. This figure corresponds to an airplane having engines supercharged up to a critical altitude h_c, above which their power decreases with increasing altitude. The units indicated on the figure are those usually employed where the English system is used. The minimum speed for level flight is almost never included in a classical performance diagram.

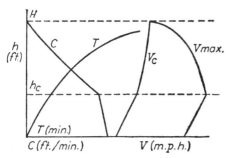

FIG. 3·6.   Classical performance diagram.

A point of technique, which greatly lessens the amount of labor required, should be mentioned. It concerns the determination of the absolute ceiling, which is almost never accomplished by actually plotting the power-available and required curves for the altitude at which they are just tangent. Instead, enough points on the rate-of-climb curve of Fig. 3·6 are determined to define it satisfactorily, and the curve is then extrapolated to $C = 0$ in order to obtain H.

3–3. Power Required Relations; Gliding Flight

In this section the power-required curve will be considered in some detail, based on the assumption that a conventional C_D vs. C_L polar is available for any airplane for which a performance estimate is desired. If a simple graphical analysis is to be used, as outlined in the last section, it is only necessary to plot curves of P_r vs. V for a series of values of σ, using the data furnished by the polar. The equations which give the transformation from polar to power-required curve follow immediately from equations 3·2 and 3·6:

$$\left.\begin{array}{l} V = \sqrt{\dfrac{2W}{\rho_0 S}} \dfrac{1}{\sqrt{C_L \sigma}} \\[4mm] P_r = VD = \tfrac{1}{2}\rho_0 S V^3 C_D \sigma = \sqrt{\dfrac{2W^3}{\rho_0 S}} \dfrac{C_D}{\sqrt{\sigma} C_L^{3/2}} \end{array}\right\} \qquad [3\cdot9]$$

The use of the English system of units in these equations will give V in feet per second and P_r in foot-pounds per second. The conversion to miles per hour and horsepower is given by

$$
\left.
\begin{array}{l}
V \text{ (m.p.h.)} = \frac{60}{88}\, V \text{ (ft./sec.)} \\[2mm]
P \text{ (hp.)} = \dfrac{P \text{ (ft. lb./sec.)}}{550}
\end{array}
\right\}
\qquad [3\cdot10]
$$

A very interesting and important conclusion which can be drawn from equation $3\cdot9$ appears if both equations are multiplied by $\sqrt{\sigma}$. The resulting right-hand sides are then entirely independent of altitude. In other words a single curve of $\sqrt{\sigma}P_r$ versus $\sqrt{\sigma}V$ completely represents the characteristics of a given airplane at all altitudes. Now $\sqrt{\sigma}V$ has a very simple physical significance, being the quantity which is shown on the dial of any Pitot tube type of air-speed indicator. A Pitot-static tube, as we saw earlier, furnishes a pressure difference, equal to the dynamic pressure $\tfrac{1}{2}\rho V^2$, which actuates the air-speed indicator hand. The dial of such an indicator has its scale divisions chosen so as to show $\sqrt{\tfrac{1}{2}\rho V^2} = \sqrt{\tfrac{1}{2}\rho_0}(\sqrt{\sigma}V)$ and (in the English system) is calibrated to read miles per hour when $\sigma = 1$. The quantity $\sqrt{\sigma}V$ which a conventional air-speed indicator exhibits, is called the "indicated air speed" and written as V_i. The true air speed at which an airplane is flying is therefore given by dividing the air-speed gage reading, V_i, by $\sqrt{\sigma}$ corresponding to the air density at the altitude in question. By analogy with this notation for velocity the conception of "indicated power" has been introduced. This is defined as $\sqrt{\sigma}$ times actual power and is also denoted by the subscript $(\ )_i$. Accordingly

$$
\left.
\begin{array}{l}
V_i = \sqrt{\sigma}V = \text{Indicated air speed} \\[2mm]
P_{r_i} = \sqrt{\sigma}P_r = \text{Indicated power required}
\end{array}
\right\}
\qquad [3\cdot11]
$$

The complete power-required characteristics of any airplane for all altitudes are given by a single curve of indicated power required vs. indicated velocity as shown in Fig. $3\cdot7$. For standard, sea-level conditions this curve is identical with the true power vs. velocity curve.

Graphical methods of analysis are always tedious, their accuracy is limited, and they often obscure the essential physical relationships involved in a problem. These difficulties very early became apparent to aerodynamicists working with performance problems, and many attempts were accordingly made to develop analytical methods in place of the graphical ones outlined above. The most successful and widely used of these have been worked out over a period of years at the GALCIT * and will be briefly discussed here. They are fundamentally based on the conception of approximating to the airplane polar by a parabola as is suggested by the Prandtl wing theory. It will be recalled that if for an elliptical wing the profile drag coefficient is constant, then the polar is a parabola of the form $C_D = a + bC_L^2$, where $b = 1/(\pi \mathcal{R})$. This sug-

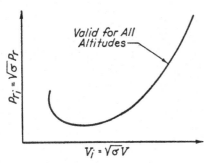

gests that the attempt be made to fit the polars of complete airplanes by a similar curve over the normal flying range, say from $C_L = 0.1$ to 0.8 or 1.0. Such attempts prove to be extremely successful in practically all cases, the deviation of the parabola from the polar in the region in question being quite small. The usual procedure in fitting is to require the parabola to coincide with the polar at two points (for example, at $C_L = 0.25$ and 0.70) which determines the coefficients a and b. The parabola is then written as

FIG. 3·7. Indicated power-required curve.

$$C_D = C_{D_{p_e}} + \frac{C_L^2}{\pi \mathcal{R}_e} \qquad [3·12]$$

where

$$\mathcal{R}_e = e \frac{(kb)^2}{S} = \text{Effective aspect ratio}$$

$$e = \text{Airplane efficiency factor (Oswald)}$$

$$C_{D_{p_e}} = \text{Effective parasite drag coefficient}$$

In this way the airplane is replaced by a single equivalent wing having elliptic lift distribution, constant profile drag coefficient, and aspect ratio equal to e times the aspect ratio which would be calculated from the wing cellule geometry. It should be noted that the approximating parabola is completely defined by two aerodynamic parameters, $C_{D_{p_e}}$

* C. B. Millikan, "The Induced Drag Viewpoint of Performance," *Aviation*. August 17, 1929.

W. B. Oswald, "General Formulas and Charts for the Calculation of Airplane Performance," *N.A.C.A. Tech. Rep.* 408 (1932).

R. J. White and V. J. Martin, "Charts for Calculating the Performance of Airplanes Having Constant Speed Propellers," *N.A.C.A. Tech. Note* 579 (1936).

W. C. Rockefeller, "General Airplane Performance," *N.A.C.A. Tech. Rep.* 654 (1939).

and $\mathcal{R}_e$, both of which can be evaluated from the polar. Actually $(kb)^2/S$ can be calculated from the airplane geometry so that $C_{D_{p_e}}$ and e are the two parameters which must be determined empirically from the polar, or otherwise estimated by the aerodynamicist. Figure 3·8 shows a typical airplane polar, the approximating parabola, and the values of $C_{D_{p_e}}$ and $\mathcal{R}_e$ deduced from the parabola.

Before considering the power required curve in the light of this approximation two important consequences of the parabolic polar should

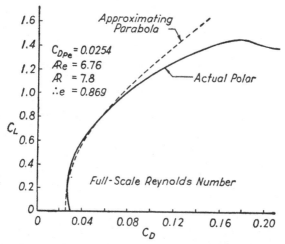

Fɪɢ. 3·8. Airplane polar and the approximating parabola. (*The difference between polar and parabola is here larger than usual.*)

be noted. The first is connected with the maximum value of the L/D ratio, whose significance can be seen if $T = 0$ is put in equations 3·1, thereby restricting the considerations to power-off or gliding flight. Equations 3·1 may now be written

$$L = W \cos \theta$$

$$D = - W \sin \theta$$

Dividing the first by the second we obtain

$$L/D = - \cot \theta$$

so that the L/D ratio determines the power-off flight path angle or angle of glide, and $(L/D)_{\max}$ corresponds to the minimum angle of glide or the flattest gliding path. This is a characteristic of great practical importance and warrants some study, which is most conveniently accomplished by considering $\left(\dfrac{D}{L}\right)_{\min}$ rather than $\left(\dfrac{L}{D}\right)_{\max}$. From equation $3\cdot12$

$$\frac{D}{L} = \frac{C_D}{C_L} = \frac{C_{D_{p_e}}}{C_L} + \frac{C_L}{\pi \mathcal{R}_e} \qquad [3\cdot13]$$

The value of C_L at which D/L has its minimum value is obtained by equating to zero the derivative of the expression on the right with respect to C_L. Carrying out the differentiation we have

$$-\frac{C_{D_{p_e}}}{C_{L_{ld}}^2} + \frac{1}{\pi \mathcal{R}_e} = 0$$

Therefore

$$C_{L_{ld}} = \sqrt{\pi}\,\sqrt{C_{D_{p_e}}\mathcal{R}_e} \qquad [3\cdot14]$$

where the subscript $(\ )_{ld}$ denotes the condition of $(L/D)_{\max}$. By substituting equation $3\cdot14$ in $3\cdot13$ and by inverting we obtain

$$\left(\frac{L}{D}\right)_{\max} = \frac{\sqrt{\pi}}{2}\sqrt{\frac{\mathcal{R}_e}{C_{D_{p_e}}}} \qquad [3\cdot15]$$

It is interesting to note that

$$C_{D_{p_e}} = \frac{C_{L_{ld}}^2}{\pi \mathcal{R}_e} = (C_{D_{i_e}})_{ld} \qquad [3\cdot16]$$

where

$$C_{D_{i_e}} = \text{the "effective" induced drag coefficient}$$

so that the maximum value of L/D occurs at the attitude where the parasite and induced drags are just equal.

The second consequence of a parabolic polar is connected with the condition of minimum power required. This has an obvious practical importance, giving the minimum horsepower which must be furnished by the power plant in order that level flight may be maintained. For a given airplane at a definite altitude equations $3\cdot9$ give

$$P_{r_{\min}} = \sqrt{\frac{2W^3}{\rho_0 S}}\,\frac{1}{\sqrt{\sigma}}\left(\frac{C_{D_{p_e}}}{C_L^{\frac{3}{2}}} + \frac{C_L^{\frac{1}{2}}}{\pi \mathcal{R}_e}\right)_{\min} \sim \left(\frac{C_{D_{p_e}}}{C_L^{\frac{3}{2}}} + \frac{C_L^{\frac{1}{2}}}{\pi \mathcal{R}_e}\right)_{\min} \qquad [3\cdot17]$$

The condition of minimum power-required, denoted by the subscript $(\ )_{mp}$, is determined by setting the derivative of equation $3\cdot17$ with respect to C_L equal to zero. This gives

$$-\frac{3}{2}\frac{C_{D_{p_e}}}{C_{L_{mp}}^{5/2}} + \frac{1}{2}\frac{1}{\pi\,\mathcal{R}_e C_{L_{mp}}^{1/2}} = 0$$

Therefore

$$C_{L_{mp}}^2 = 3\pi C_{D_{p_e}}\mathcal{R}_e$$

Therefore

$$C_{L_{mp}} = \sqrt{3\pi}\,\sqrt{C_{D_{p_e}}\mathcal{R}_e} = \sqrt{3}C_{L_{ld}} \qquad [3\cdot18]$$

Substituting this in equation $3\cdot17$ we obtain

$$P_{r_{min}} = \frac{4\sqrt{2}}{(3\pi)^{3/4}}\sqrt{\frac{W^3}{\rho_0 S}}\left(\frac{C_{D_{p_e}}}{\sigma^2\mathcal{R}_e^3}\right)^{1/4} \qquad [3\cdot19]$$

Writing as before $C_{D_{i_e}} = \dfrac{C_L^2}{\pi\,\mathcal{R}_e}$ we have from equation $3\cdot18$

$$(C_{D_{i_e}})_{mp} = \frac{3\pi C_{D_{p_e}}\mathcal{R}_e}{\pi\,\mathcal{R}_e} = 3C_{D_{p_e}} \qquad [3\cdot20]$$

so that at the condition of minimum power required the induced drag is just three times as large as the parasite drag.

The above relations may be considered as examples of the way in which the assumption of a parabolic polar leads to definite analytical results in the performance problem. A more systematic treatment of the power-required expression using this assumption follows:

$$P_r = VD = C_D\frac{\rho}{2}SV^3 = \frac{\rho_0}{2}S\,(C_{D_{p_e}} + C_{D_{i_e}})\sigma V^3$$

Now

$$C_{D_{i_e}} = \frac{C_L^2}{\pi\,\mathcal{R}_e} = \frac{4W^2}{\pi\,\mathcal{R}_e\rho^2 S^2 V^4} = \frac{4W^2}{\pi\rho_0^2\,\mathcal{R}_e S^2\sigma^2 V^4}$$

using equation $3\cdot2$. Hence

$$P_r = \frac{\rho_0}{2}C_{D_{p_e}}S\sigma V^3 + \frac{2}{\pi\rho_0}\frac{W^2}{\mathcal{R}_e S}\frac{1}{\sigma V}$$

It is convenient to introduce

$$\left.\begin{array}{l} b_e^2 = e(kb_1)^2 = \mathcal{R}_e S = \text{(Effective span)}^2 \\[4pt] f = C_{D_{p_e}}S = \text{Equivalent parasite area} \end{array}\right\} \qquad [3\cdot21]$$

in terms of which the power required becomes

$$P_r = \frac{\rho_0 f}{2}\sigma V^3 + \frac{2}{\pi\rho_0}\frac{W^2}{b_e^2}\frac{1}{\sigma V} \qquad [3\cdot22]$$

In a later analysis of the complete performance equation $3 \cdot 6$ it will be found convenient to divide all of the power terms by W, which leads to terms having the dimensions of velocity. The term P_r/W in particular gives the so-called sinking speed of an airplane when the plane is gliding without power. This appears at once if equation $3 \cdot 6$ is divided by W and P_a is put equal to zero. Then $P_r/W = - \, dh/dt$ = the rate at which altitude is lost. Accordingly we define

$$w_s = \frac{P_r}{W} = \text{Sinking speed} \qquad [3 \cdot 23]$$

Then from equation $3 \cdot 22$

$$w_s = \frac{\rho_0}{2} \frac{f}{W} \sigma V^3 + \frac{2}{\pi \rho_0} \frac{W}{b_e^2} \frac{1}{\sigma V} \qquad [3 \cdot 24]$$

This equation suggests the introduction of two new airplane parameters which later are seen to be fundamental to the whole performance analysis. Two forms of these parameters are convenient, the first, called physical, for use in theoretical analysis, and the second, called engineering, for numerical performance computations. The definitions, dimensions, and units in the English system, are collected below:

$$\left.\begin{array}{l}
\lambda_p = \dfrac{2W}{\rho_0 f} = \text{Parasite loading (physical); dimensions } |V^2|; \\
\qquad\qquad \text{units (ft./sec.)}^2 \\[2mm]
\lambda_s = \dfrac{2W}{\pi \rho_0 b_e^2} = \text{Span loading (physical); dimensions } |V^2|; \\
\qquad\qquad \text{units (ft./sec.)}^2 \\[2mm]
l_p = \dfrac{W}{f} = \text{Parasite loading (engineering); dimensions} \\
\qquad\qquad |\text{ pressure }|; \text{ units (lb./ft.}^2) \\[2mm]
l_s = \dfrac{W}{b_e^2} = \text{Span loading (engineering); dimensions} \\
\qquad\qquad |\text{ pressure }|; \text{ units (lb./ft.}^2)
\end{array}\right\} \quad [3 \cdot 25]$$

$$\lambda_p = 841.0 \, l_p; \quad \lambda_s = 267.7 \, l_s$$

In terms of these parameters equation $3 \cdot 24$ assumes the simple form

$$w_s = \frac{P_r}{W} = \frac{\sigma V^3}{\lambda_p} + \frac{\lambda s}{\sigma V} = \frac{\sigma V^3}{841 l_p} + \frac{267.7 l_s}{\sigma V} \qquad [3 \cdot 26]$$

Equation $3 \cdot 26$ permits the rapid calculation of sinking speed or power required as function of V for any airplane (given by W, l_p, l_s) at any altitude (σ). A single equation valid for any altitude replaces $3 \cdot 26$ if the following "indicated" quantities are introduced:

$$V_i = \sqrt{\sigma} V, \qquad P_{r_i} = \sqrt{\sigma} P_r, \qquad w_{s_i} = \sqrt{\sigma} w_s \qquad [3 \cdot 27]$$

By multiplying equation 3·26 by $\sqrt{\sigma}$ the simplest form of the power required-velocity relation is obtained immediately:

$$w_{s_i} = \frac{P_{r_i}}{W} = \frac{V_i^3}{\lambda_p} + \frac{\lambda_s}{V_i} = \frac{V_i^3}{841 l_p} + \frac{267.7 l_s}{V_i} \qquad [3\cdot28]$$

In addition to furnishing a convenient formula for calculating the coordinates of a power-required curve, equation 3·28 also affords a considerable amount of insight into the nature of such a curve. Figure 3·9 shows a P_{r_i} curve for a typical airplane as calculated from equation 3·28. The two additive terms which make up the total P_{r_i} are also indicated, that corresponding to the second term of the equation being

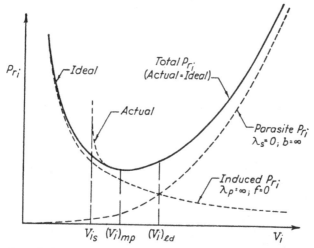

FIG. 3·9.   Indicated power-required curve showing induced and parasite contributions.

labeled "induced P_{r_i}" and that corresponding to the first term being called "parasite P_{r_i}." The reason for the terminology is obvious in view of the type of drag which corresponds to each term. The points corresponding to maximum L/D and minimum P_{r_i} are shown on the figure. A very important characteristic is also brought out by a comparison of the portions of the curve labeled "ideal" and "actual." The "ideal" curve corresponds exactly to equation 3·28 and therefore rests on the assumption that V may go to zero, i.e., C_L to ∞. The polar for an actual airplane breaks away from the parabola on which equation 3·28 is based as C_{Lmax} is approached, so that the actual P_{r_i} curve must also break away from the "ideal" one as V_i approaches the indicated stalling speed, V_{i_s}. For most airplanes the deviation between the "ideal" and "actual" curves becomes appreciable as V_i becomes smaller than $V_{i_{mp}}$. Since for normal performance problems we are not concerned with this region of small velocities the deviation is seldom of practical significance.

A further discussion of equation $3 \cdot 28$ will be found in section 3–5 after consideration of the power-available aspect of the performance problem, which is necessary before the analysis can be developed further.

3–4. Power Available; Engine and Propeller Characteristics

The discussion of power available given in this chapter is necessarily extremely limited in scope. It will therefore be restricted to airplanes equipped with constant speed propellers, i.e., where the propeller revolutions per minute can be set and maintained by the pilot at any fixed value desired. Unless voluntary changes are made, therefore, the number of revolutions per minute does not vary with either V or σ but may be considered as a predetermined constant. The analysis of power available when independent revolution control does not exist is more complex although not different in principle and, since practically all modern high performance airplanes are equipped with constant-speed propellers, its omission is not of much practical importance.

The engine charts furnished with modern engines give brake horse-power, P, delivered to the propeller, as a function of revolutions per minute, N, manifold pressure, MP, and standard altitude, h. Figure $3 \cdot 10$ shows such an engine chart with an illustrative example. In general, with constant-speed propellers, performance estimates are required at constant MP below critical altitude, at full throttle above critical altitude, or at some preassigned constant power. Since the revolution speed is also preassigned it follows that both P and N are known functions of h (or σ) which are independent of airspeed V. This is the great simplification involved in the use of constant-speed propellers since it makes any further discussion of engine characteristics unnecessary for our present purpose, all that is needed for a performance estimate being the operating procedure imposed on the engine, and the engine chart. From these two are obtained

$$P(\sigma) \quad \text{and} \quad N(\sigma) \quad \text{are given functions} \qquad [3\cdot29]$$

A discussion of propeller characteristics involves the use of a rather special terminology and notation which will be developed. Consider the element of a propeller blade just at the tip and imagine the helical sur-

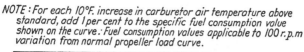

NOTE: For each 10°F. increase in carburetor air temperature above standard, add 1 per cent to the specific fuel consumption value shown on the curve: Fuel consumption values applicable to 100 r.p.m variation from normal propeller load curve.

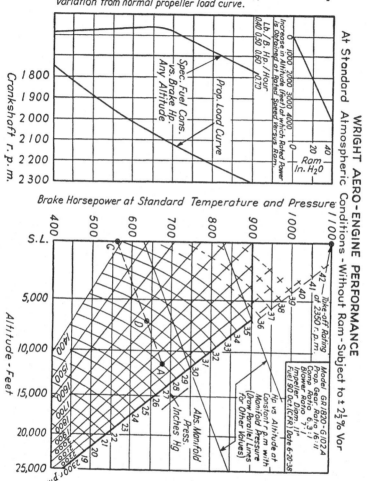

Courtesy of Wright Aeronautical Corp.

FIG. 3·10. Engine chart for Wright Cyclone Engine.

Numerical example: Suppose manifold pressure is 28 in. at 2100 r.p.m. at an altitude of 7000 ft. Locate point *A* on full-throttle curves at 28 in. and 2100 r.p.m.; draw a line *AC* through *A* parallel to the typical lines provided. Line *AC* denotes brake horsepower versus altitude at 28 in. and 2100 r.p.m.; thus, at 7000 ft. the engine develops 640 b.hp. (point *D*).

face, on which this travels, unwrapped onto a plane. Then Fig. 3·11, representing the blade element on this plane, indicates the motion of the element. V, n, and d are forward velocity, revolution speed, and propeller diameter expressed in consistent units.* The horizontal vector represents the velocity component due to rotation; the vertical vector, that due to the forward motion of the propeller; and the hypotenuse vector, the resultant velocity of the element. The angle β_t is the blade setting of the (dotted) chord line of the element. It is apparent that

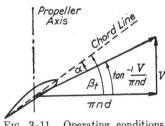

the angle of attack, α, of the element is determined by the blade setting and $\tan^{-1} V/\pi nd$. For a fixed-pitch propeller β_t is constant, so that the variable defining the angle of attack, and hence the operating conditions of the blade element, is $V/\pi nd$. For each blade element a similar diagram could be drawn giving the operating conditions for the element in question. However, for any fixed value of $V/\pi nd$ the distri-

FIG. 3·11. Operating conditions for propeller tip blade element.

bution of α along the blade is determined by the variation in blade angle or twist which is built into the propeller. Hence the value of $V/\pi nd$, or V/nd, and the blade setting of any one blade element define the operating conditions of any propeller as a whole, whether it be of the fixed or adjustable pitch type. In practice the blade angle β for a propeller is almost always measured at the station three-fourths of the distance from the propeller axis to the tip. The ratio V/nd is often called the advance ratio and denoted by the symbol J. Hence the geometrical conditions under which any member of a family of similar propellers is working are given by the values of the two dimensionless parameters β and J.

The propeller characteristics important in performance studies are the brake horsepower absorbed, P, the forward thrust, T, and the thrust power delivered, $TV = P\eta$, where η is the efficiency. Dimensional con-

---

* In accordance with a widely used convention, both symbols N and n for revolution speed will be used with the following definitions:

　n = propeller revolutions in unit time where the time unit is the basic one in the consistent set of units employed in the analysis, e g., n = propeller revolutions per second in the ft., lb., sec. system.

　N = propeller or engine revolutions in unit time where this unit is the usual one of conventional engineering practice. e.g., N = revolutions per minute in most English-speaking countries.

siderations show that power and thrust may be written in the form

$$P = C_k J^k \rho n^3 d^5$$
$$T = C_l J^l \rho n^2 d^4 \qquad \bigg\} \qquad [3\cdot30]$$

where C_k and C_l are dimensionless coefficients and k and l are arbitrary numbers. C_k and C_l are the dimensionless quantities which depend on the geometrical conditions of operation of a propeller and hence, for a family of similar propellers, are functions of the dimensionless operating parameters β and J.

By giving k and l simple integral values and solving equations $3\cdot30$ for C_k and C_l, it is clear that these coefficients may be expressed in terms of P, ρ, and any two of the quantities V, n, and d. For different purposes it is convenient to choose different values of k and l, thus defining different pairs of coefficients C_k and C_l. The most generally used coefficients correspond to $k = l = 0$ and are given special symbols:

$$C_P = P/(\rho n^3 d^5) = \text{Power coefficient}$$
$$C_T = T/(\rho n^2 d^4) = \text{Thrust coefficient} \qquad \bigg\} \qquad [3\cdot31]$$

Many other coefficients based on other values of k and l are used in various propeller problems, but these two will be sufficient for our purposes. The efficiency bears a close relation to these coefficients for

$$\eta = \frac{TV}{P} = \frac{C_T}{C_p} \frac{V}{nd} = \frac{C_T J}{C_p} \qquad [3\cdot32]$$

All of the dimensionless quantities defined in equations $3\cdot31$ and $3\cdot32$ are experimentally determined functions of β and J for a particular family of propellers, as was indicated following equation $3\cdot30$.

Both the power absorbed and the power delivered by a propeller are required for a performance analysis, so that two coefficients must be given. The two propeller charts given in Figs. $3\cdot12$ and $3\cdot13$ furnish such data in a convenient form for a widely used three-bladed family of propellers investigated by the N.A.C.A.

By combining the engine and propeller characteristics we obtain

$$P_a = P\eta = F(V, \sigma) \qquad [3\cdot33]$$

The function $F(V, \sigma)$ gives the dependence of power available on flight velocity and density, as is necessary for the plotting of power available in a performance-power diagram. In determining this function $P(\sigma)$ and $N(\sigma)$ are given from the engine operating data. It is also assumed that the propeller diameter d is known. Then for a given altitude, σ, C_P is a constant which may be calculated from the convenient engineering formula

$$C_P = 0.5 \frac{(\text{b.hp.}/1000)}{\sigma (\text{r.p.m.}/1000)^3 (\text{dia. in ft.}/10)^5} \qquad [3\cdot34]$$

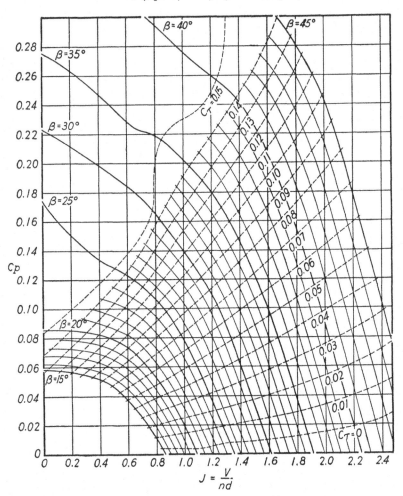

Fig. 3·12. Three-bladed propeller chart.

J depends only on the velocity V, being given by

$$J = 88 \frac{V \text{ in m.p.h.}}{(\text{r.p.m.})(\text{dia. in ft.})} \qquad [3\cdot35]$$

Equation 3·35 permits J to be determined as a function of V for a definite altitude corresponding to a determined value of N. The propeller charts together with these results give $\eta(V)$ for the altitude in question so that $P_a(V)$ is determined. The process may be repeated for any other altitudes desired so that finally curves of P_a vs. V, or of $P_{a_i} = \sqrt{\sigma}P_a$ vs. V_i can be drawn on the performance-power diagram for all of the altitudes required. The result is a family of power-available curves like that of Fig. 3·5, or an analogous family of indicated power-available curves plotted on a diagram with the coordinates of Fig. 3·7.

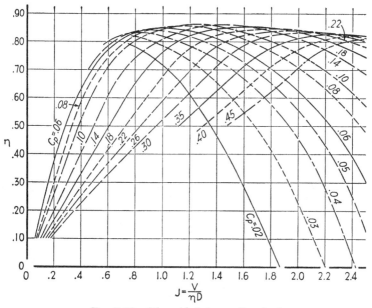

FIG. 3·13.   Three-bladed propeller chart.

In this connection it should be pointed out that the use of indicated coordinates does not reduce the number of power-available curves which is necessary, as it does for power-required curves. In fact, the spread between the indicated power-available curves for a series of altitudes is greater than that for the corresponding true power-available curves.

3–5. Power Required and Available Combined; Analysis of the Classical Performance Problem

The fundamental performance equation 3·6 is written in the form

$$W\frac{dh}{dt} = AP\eta - P_r \qquad [3\cdot36]$$

The numerical conversion factor, A, has been introduced here in order that the brake horsepower, P, may be expressed in horsepower units while P_r and $W(dh/dt)$ are in the usual physical units. In the English system we have

$$A = 550$$

In section 3–3 it was indicated that it is convenient to divide equation 3·36 by W, which gives

$$\frac{dh}{dt} = \frac{AP\eta}{W} - \frac{P_r}{W} = \frac{AP\eta}{W} - \frac{\sigma V^3}{\lambda_p} - \frac{\lambda_s}{\sigma V} \qquad [3\cdot37]$$

introducing equation 3·26. The last two terms contain the variables σ and V and also the fixed "design parameters" λ_p and λ_s. This suggests the desirability of putting the first term on the right into a similar form by separating it into two parts, one a design constant and the other giving the variation with σ and V. Accordingly we write

$$\frac{AP\eta}{W} = \frac{AP_0\eta_0}{W} R_P$$

where

$$\left. \begin{aligned} P_0\eta_0 &= \text{Design thrust horsepower} \\ R_P = \frac{P\eta}{P_0\eta_0} &= \text{Power available ratio} \end{aligned} \right\} \qquad [3\cdot38]$$

The design thrust horsepower is the power available under certain specified design conditions. These often correspond to full rated power of the engine and a propeller efficiency which occurs at maximum level speed at critical altitude. Whatever the specified design conditions, $P_0\eta_0$ may be considered as a design parameter of the airplane independent of V and σ. The dimensionless ratio R_P is, however, a function of V and σ and gives the variation of power available with these variables. A thrust power loading analogous to the span and parasite loadings is next introduced. The physical and engineering forms are defined as follows:

$$\left. \begin{aligned} \lambda_t &= \frac{W}{AP_0\eta_0} = \text{Thrust power loading (physical); dimensions} \\ &\qquad |1/V|; \text{ units (sec./ft.)} \\ l_t &= \frac{W}{P_0\eta_0} = \text{Thrust power loading (engineering); dimensions } |1/V|; \text{ units (lb./hp.)} \end{aligned} \right\} \qquad [3\cdot39]$$

With this notation we have

$$\frac{P_a}{W} = \frac{AP\eta}{W} = \frac{R_P}{\lambda_t} \qquad [3\cdot40]$$

and the performance equation 3·37 becomes

$$\frac{dh}{dt} = \frac{R_P(\sigma, V)}{\lambda_t} - \frac{\sigma V^3}{\lambda_p} - \frac{\lambda_s}{\sigma V} \qquad [3\cdot41]$$

where the dependence of R_P on σ and V has been explicitly indicated. It is convenient to put this equation into a form in which each of the additive terms is dimensionless, as can be done by multiplying through by λ_t. Then

$$\lambda_t \frac{dh}{dt} = R_P - \frac{\lambda_t}{\lambda_p} \sigma V^3 - \frac{\lambda_s \lambda_t}{\sigma V} \qquad [3\cdot42]$$

The design conditions corresponding to P_0, η_0 are now introduced explicitly. The altitude corresponding to the design brake horsepower is indicated by σ_0, and the design velocity at this altitude, which corresponds to the design propulsive efficiency, η_0, by V_0. V_0 is then defined as the level-flight velocity corresponding to the design thrust horsepower $P_0 \eta_0$. The speed ratio V/V_0 is written R_V. In terms of these quantities equation 3·42 takes the form

$$\lambda_t \frac{dh}{dt} = R_P - \frac{\lambda_t}{\lambda_p} V_0^3 \sigma R_V^3 - \frac{\lambda_s \lambda_t}{V_0} \frac{1}{\sigma R_V} \qquad [3\cdot43]$$

where

$$R_V = \frac{V}{V_0}$$

The use of "indicated" quantities greatly simplifies the consideration of the effects of variation in altitude. Therefore the following "indicated" parameters may be introduced and denoted by the subscript $(\ )_i$

$$\left.\begin{array}{l} V_i = \sqrt{\sigma} V; \quad P_i = \sqrt{\sigma} P; \quad \left(\dfrac{dh}{dt}\right)_i = \sqrt{\sigma}\, \dfrac{dh}{dt} \\[2mm] V_{i_0} = \sqrt{\sigma_0} V_0; \quad P_{i_0} = \sqrt{\sigma_0} P_0 \\[2mm] R_{V_i} = \dfrac{V_i}{V_{i_0}} = \sqrt{\dfrac{\sigma}{\sigma_0}}\, R_V; \quad R_{P_i} = \dfrac{P_i}{P_{i_0}} = \sqrt{\dfrac{\sigma}{\sigma_0}}\, R_P \\[2mm] \lambda_{t_i} = \dfrac{W}{A\sqrt{\sigma_0 P_0 \eta_0}} = \text{Indicated thrust power loading} \end{array}\right\} \qquad [3\cdot44]$$

Multiplying equation 3·43 by $\sqrt{\sigma/\sigma_0}$ the performance equation takes the form

$$\lambda_{t_i} \left(\frac{dh}{dt}\right)_i = R_{P_i} - \frac{\lambda_{t_i} V_{i_0}^3}{\lambda_p} R_{V_i}^3 - \frac{\lambda_s \lambda_{t_i}}{V_{i_0}} \frac{1}{R_{V_i}} \qquad [3\cdot45]$$

The altitude, or density, no longer appears explicitly although it is implicitly included in each term. The equation therefore contains two variables: $(dh/dt)_i$ and R_{V_i}, and three "design parameters," λ_{t_i}, $\lambda_{t_i} V_{i_0}^3/\lambda_p$, $\lambda_s \lambda_{t_i}/V_{i_0}$. R_{P_i} is a function of R_{V_i} which is assumed to be given by the engine and propeller characteristics. The last two design parameters are of great importance and, as is shown in *N.A.C.A. Technical Report* 654, have simple physical significances, so that they are represented by special symbols:

$$\left. \begin{array}{c} \Omega_i = \left(\dfrac{\lambda_{t_i}}{\lambda_p}\right)^{\frac{1}{3}} V_{i_0} \\[3mm] \Gamma_i = \dfrac{\lambda_s \lambda_{t_i}}{V_{i_0}} \end{array} \right\} \qquad [3 \cdot 46]$$

Equation 3·45 may now be written

$$\lambda_{t_i}\left(\frac{dh}{dt}\right)_i = R_{P_i} - \Omega_i^3 R_{V_i}^3 - \frac{\Gamma_i}{R_{V_i}} \qquad [3 \cdot 47]$$

It is possible to eliminate one of the three design parameters λ_{t_i}, Ω_i, Γ_i by considering level flight under the design conditions σ_0, V_0. Then

$$R_{V_i} = 1, \quad R_{P_i} = 1, \quad \left(\frac{dh}{dt}\right)_i = 0$$

and equation 3·47 becomes

$$\Gamma_i = 1 - \Omega_i^3 \qquad [3 \cdot 48]$$

Eliminating Γ_i from equation 3·47 using this relation, the performance equation becomes

$$\lambda_{t_i}\left(\frac{dh}{dt}\right)_i = R_{P_i} - \Omega_i^3 R_{V_i}^3 - \frac{1 - \Omega_i^3}{R_{V_i}} \qquad [3 \cdot 49]$$

A great simplification has here been achieved, since all level-speed characteristics, where $dh/dt = 0$, depend on a single design parameter, Ω_i, and all rates of climb depend only on this parameter and one other, λ_{t_i}, which appears as a constant of proportionality. The parameter Ω_i contains the design speed, V_{i_0}, and hence is not strictly a fundamental design parameter. It is, however, closely related to such a parameter as can be seen from the following analysis. From equation 3·46 and 3·48 it follows that

$$\Omega_i \Gamma_i = \Omega_i(1 - \Omega_i^3) = \frac{\lambda_s \lambda_{t_i}}{(\lambda_p/\lambda_{t_i})^{1/3}} = \Lambda_i' \qquad [3 \cdot 50]$$

Hence Ω_i is a definite function of the parameter Λ_i' which itself depends only on basic design characteristics of the airplane. Unfortunately the relation between Ω_i and Λ_i' is an algebraic one of the fourth degree which cannot be inverted to give an analytical expression for Ω_i in terms of Λ_i'. However, the relation can easily be plotted graphically as has been done in Fig. 3·14. Hence equation 3·49 can always be combined with equation 3·50 so that the resulting relation can be written symbolically in the form

$$\lambda_{t_i}\left(\frac{dh}{dt}\right)_i = F(\Lambda_i',\ \sigma,\ R_{V_i}) \qquad [3\cdot51]$$

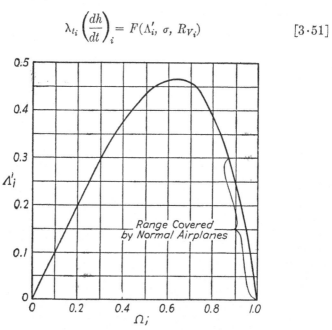

Fig. 3·14. Relation between Ω_i and the fundamental performance parameter Λ'_i.

since R_{P_i} is assumed to be a given function of σ and R_V (or R_{V_i}). In this form it appears that all level-flight characteristics depend only on the design parameter Λ_i' and all climbing characteristics on Λ_i' and λ_{t_i}. The fundamental importance of Λ_i' justifies its name: "fundamental performance parameter." The engineering form is

$$\left.\begin{aligned}\Lambda_i &= \frac{l_s l_{t_i}}{(l_p/l_{t_i})^{1/3}} = \frac{1}{\sigma^{2/3}}\frac{l_s l_t}{(l_p/l_t)^{1/3}} = \frac{1}{\sigma^{2/3}}\Lambda\\ \Lambda_i' &= 0.006293\,\Lambda_i\end{aligned}\right\} \qquad [3\cdot52]$$

The relation between Λ_i and Ω_i (corresponding to that between Λ_i' and Ω_i given in Fig. 3·14) is plotted to a large scale in the composite performance chart, Fig. 3·19 to be discussed later.

All of the parameters Λ_i', Ω_i, and Γ_i have simple and significant physical interpretations which are discussed in detail in *N.A.C.A. Technical Report* 654.* Unfortunately only an outline can be given here of the procedure for actually calculating performance using these parameters. Assume, therefore, a definite type of propulsive system so that R_P is a given function of σ and R_{V_i}, and the performance equation is symbolically represented by equation 3·51.

The first step in such a calculation is the determination of the loadings l_s, l_p, and l_{t_i}. Then the quantities $l_s l_{t_i}$, $(l_p/l_{t_i})^{1/3}$, and Λ_i are determined. From Fig. 3·14 or Fig. 3·19 (the curve corresponding to equation 3·50) Ω_i is obtained. Then from equation 3·46

$$V_{i_0} = (\lambda_p/\lambda_{t_i})^{1/3}\Omega_i$$

or in engineering units

$$V_{i_0} \text{ (m.p.h.)} = 52.73\ (l_p/l_{t_i})^{1/3}\Omega_i = \sqrt{\sigma_0}V_0 \text{ (m.p.h.)} \qquad [3\cdot53]$$

The basic design velocity is thus known, and all later velocity determinations can be made in the form of speed ratios with V_{i_0} constituting the denominator of the ratio. One difficulty which has not been pointed out in the above procedure is the fact that η_0 must be known in order to calculate $l_{t_i} = W/\sqrt{\sigma_0}P_0\eta_0$. But η_0 is determined on a propeller chart from the values of C_P and J corresponding to the "design" conditions. P_0, N_0, σ_0, and d are assumed to be known so that $C_{P_0} = C_P$ (design conditions) is given. However, $J_0 = V_0/n_0d$ cannot be determined until V_0 is known. A method of successive approximations is therefore used: An initial value of η_0 is assumed, and V_0 determined as outlined above; this determines a first value of J_0 corresponding to which an η_0 is determined from the propeller charts; if this does not agree with the assumed η_0, the above process is repeated using the new η_0. This is continued until the assumed and determined η_0's check. With a little experience a single repetition will almost always suffice to give the final η_0 and V_0.

For maximum level speed at altitudes other than that corresponding to σ_0, $(dh/dt)_i$ is put equal to zero in the performance equation 3·49. In view of equation 3·50 this leads to an equation of the form

$$F_1(\Lambda_i, \sigma, R_{V_{i_{max}}}) = 0 \qquad [3\cdot54]$$

which can be solved graphically to give $R_{V_{i_{max}}}$ as function of Λ_i for constant values of σ. The solution will of course depend on the nature of the power-available function $R_{P_i}(\sigma, R_{V_i})$. Oswald † has carried out

* In this report the subscript σ is used in place of i and is attached to λ_s and λ_p rather than to λ_t. The physical discussion is, however, unaffected by these differences.

† *Loc. cit.*

the calculations and given charts for the case of unsupercharged engines with certain fixed-pitch propellers, White and Martin* for unsupercharged engines with a definite family of constant-speed propellers, and

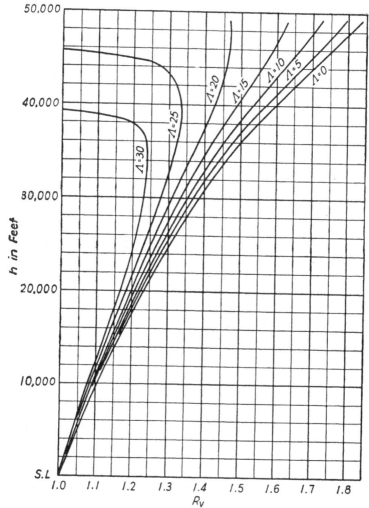

FIG. 3·15.   Maximum level speed ratio at altitude for constant power available.

Rockefeller* for the idealized case in which $R_P = 1$; i.e., $P\eta$ does not vary with σ or V, and the design altitude (σ_0) is taken as sea level. Figure 3·15 shows Rockefeller's results in a different form than that

* *Loc. cit.*

presented in *N.A.C.A. Technical Report* 654. It indicates the great gains in speed which would follow from flying at high altitudes if the power available could be kept constant as the altitude was increased.

Because modern propulsive systems do not usually have the characteristics assumed by the above authors the procedure recommended for finding the velocity at several altitudes is to consider each altitude in turn as a "design altitude," write the corresponding power as P_0, and calculate the velocity for each altitude independently by the procedure outlined above for V_0.

Maximum rate of climb and the corresponding speed, V_c, can be determined by differentiating equation 3·49 with respect to R_{V_i} and equating to zero. This leads to a relation of the form

$$F_2(\Lambda_i', \sigma, R_{V_{c_i}}) = 0 \qquad [3·55]$$

where $R_{V_{c_i}}$ = indicated speed ratio for best climb for an airplane corresponding to Λ_i' at an altitude σ. This is of the same form as equation 3·54 and can be solved and the results plotted in the same way for any given function $R_{P_i}(\sigma, R_{V_i})$. If the value of $R_{V_{c_i}}$ determined in this way is substituted back into the performance equation 3·49 the result is an expression which may be indicated by

where
$$\left.\begin{array}{l} \lambda_{t_i} C_i = F_3(\Lambda_i', \sigma) \\[2mm] C_i = (dh/dt)_{i\max} \text{ for the } \sigma \text{ in question} \end{array}\right\} \qquad [3·56]$$

The appropriate charts have been presented by Oswald and White-Martin for the special cases which they considered.

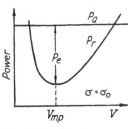

Rockefeller has given a procedure which, although more lengthy than those of the above authors, is much more general. In this he first considers the case

$$R_P = 1, \qquad \sigma = \sigma_0$$

i.e., he considers the conditions at critical altitude for an airplane with constant power available (cf. Fig. 3·16). For this case $V_c = V_{mp}$ and from equations 3·44

FIG. 3·16. Power curves for an airplane with constant power available, at critical altitude.

$$R_{V_i} = R_V; \quad \lambda_{t_i} C_i = \frac{\lambda_t}{\sqrt{\sigma}} \sqrt{\sigma} C = \lambda_t C$$

Furthermore σ does not appear explicitly in equations 3·55 and 3·56 since σ enters these equations only through R_P which, here, is constant.

Hence, using the subscript $(\ )_{mp}$ instead of $(\ )_c$ to indicate that the conditions correspond to minimum power required, these equations take the form, in engineering units,

$$\left.\begin{array}{l} F_2(\Lambda_i, R_{V_{mp}}) = 0 \\[2mm] l_t C_{mp} = F_3(\Lambda_i) \end{array}\right\} \qquad [3\cdot57]$$

These two relations have been worked out numerically by Rockefeller and are presented graphically in the composite performance chart, Fig. 3·19.

Unfortunately no actual airplane has a power available which is independent of speed so that the results are not immediately applicable

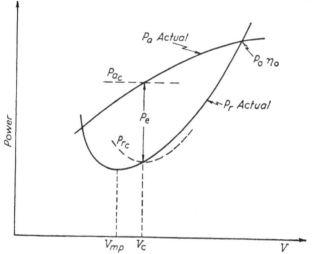

Fɪɢ. 3·17. Power curves for an actual airplane and the associated "climbing" airplane.

to the problem of maximum rate of climb. Rockefeller's solution of the problem is illustrated in Fig. 3·17. He considers a fictitious or "climbing" airplane, having constant power available, whose maximum rate of climb and speed for best climb are identical with the true values for the actual airplane. The subscript $(\ )_c$ is used for this fictitious "climbing" airplane. The following notation is introduced.

$$\left.\begin{array}{l} R_R = \dfrac{V_c}{V_{mp}} \\[4mm] l_{t_c} = \dfrac{W}{P_c \eta_c} = l_t \dfrac{P_0 \eta_0}{P_c \eta_c} = l_t \dfrac{\eta_0}{\eta_c} \text{ for constant speed propellers} \end{array}\right\} \qquad [3\cdot58]$$

R_R has been empirically determined as a function of Λ_i, the relation being given in Fig. 3·19. Rockefeller proves that the Λ_{i_c} corresponding to the P_a and P_r curves of the fictitious climbing airplane is given by

$$\Lambda_{i_c} = R_\Lambda \left(\frac{P_0 \eta_0}{P_c \eta_c}\right)^{\!\!4/3} \Lambda_i = R_\Lambda \left(\frac{\eta_0}{\eta_c}\right)^{\!\!4/3} \Lambda_i \qquad \begin{array}{l}\text{for constant-speed}\\\text{propellers}\end{array} \qquad [3·59]$$

where R_Λ is a function of R_R which is given in Fig. 3·19. Corresponding to this Λ_{i_c} the value of $(l_t C_{mp})_c$ is given by the $l_t C_{mp}$ vs. Λ_i curve already discussed. Since $(C_{mp})_c \equiv C$, the true rate of climb is obtained by dividing $(l_t C_{mp})_c$ by l_{t_c}.

The steps followed in determining C for the altitude corresponding to σ_0 may be recapitulated as follows for constant-speed propellers:

(a) Having Λ_i look up R_R and $R_{V_{mp}}$ on the composite performance chart, Fig. 3·19.

(b) Compute $R_{V_c} = R_R R_{V_{mp}}$.

(c) Compute $J_c = R_{V_c} J_0$ and determine $\eta_c(C_{P_0}, J_c)$ from the propeller chart.

(d) Look up $R_\Lambda(R_R)$ on the composite performance chart.

(e) Compute $l_{t_c} = l_t \eta_0 / \eta_c$ and $\Lambda_{i_c} = R_\Lambda (\eta_0/\eta_c)^{4/3} \Lambda_i$

(f) Determine $l_t C_{mp}$ corresponding to Λ_{i_c} on the composite performance chart.

(g) Determine C by dividing this $l_t C_{mp}$ by l_{t_c}.

(h) Determine $V_c = R_{V_c} V_0$.

The only empirical element in this analysis is the curve giving R_R as a function of Λ_i. The method gives exactly the true rate of climb corresponding to the velocity $V_c = R_R R_{V_{mp}} V_0$ whether or not this is actually the speed for best climb. If R_R is not accurate the rate of climb determined will not be quite the true maximum. This can be verified by repeating the calculations for other values of R_R differing a little from that first used. In general, however, the empirical curve for R_R is sufficiently accurate so that the corresponding rate of climb differs negligibly from the true maximum value. The above procedure can be repeated for other altitudes by choosing a different σ_0, exactly as was done in finding $V_{\max}(\sigma)$.

One of the very important consequences of the analytic performance method outlined above is that it furnishes an accurate method of predicting the change in performance which will result from a small change in the basic design constants of an airplane: W, f, b, and P. Such problems continually occur in connection with the design of a new airplane or the modification of an existing one. The analysis leading to the final results is too elaborate to discuss here, but is given in full in *N.A.C.A. Technical Report* 654. The results themselves are contained in Fig. 3·18 which gives the percentage change in V_{max}, C, and H corresponding to a 1 per cent change in the basic parameters. All relations depend only on the original value of Λ_i before the modification. This figure has proven to be extremely valuable in practice.

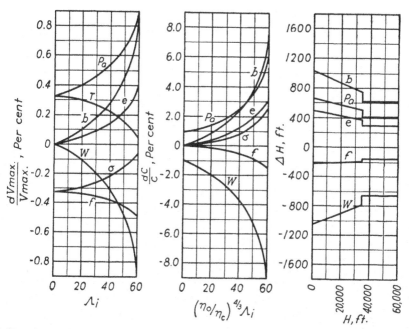

FIG. 3·18.   Charts giving variation in performance corresponding to a one per cent change in airplane parameters.

The procedure for completing the classical performance problem has been outlined and the section will be closed with a few remarks about the fundamental performance parameter Λ_i. It is easy to see that the lower the value of Λ_i, the higher is the over-all performance of an airplane. Conventional modern airplanes at moderate altitudes have values of Λ_i which usually lie between about 5 and 20. It is therefore very easy to make a first estimate of the general performance characteristics of an airplane solely from the size of the Λ_i associated with it. The exact analysis shows that there is a definite upper limit to the possible value of Λ_i at approximately 75. In fact an airplane whose Λ_i at a given altitude is 75 is actually flying at its absolute ceiling. No airplane can possibly fly under conditions where its Λ_i exceeds 75. This definite numerical limit is one of the most remarkable features which this method of performance analysis furnishes.

As a result of much experience at the GALCIT a composite performance chart has been developed which includes all of the curves required in the solution of the classical performance problem using Rockefeller's method. This chart is reproduced to a small scale in Fig. 3·19. A

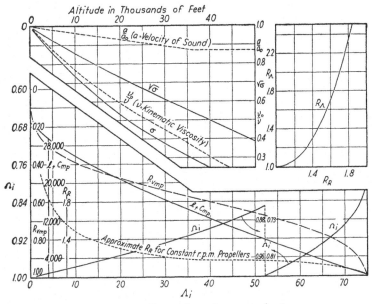

Fig. 3·19.   Composite performance chart.

large scale version, suitable for actual computation purposes and outlining the steps to be followed in a performance calculation, is given in Chart B. A complete solution of the classical performance problem may be carried out using only Chart B, propeller curves, an engine chart and a slide-rule.

3–6. Special Performance Problems

(a) *Range and endurance.* The determination of range and endurance would be relatively easy if propulsive efficiency and specific fuel consumption (c = pounds of fuel consumed per brake horsepower per hour) were constant, i.e., independent of flight velocity or power delivered by the engine. Unfortunately neither is accurately constant, but within the range of interest their variation is usually small enough so that a fairly satisfactory first approximation can be obtained by assuming average constant values.   With these assumptions we have

(1) Maximum range $\sim$ minimum pounds of fuel consumed per mile $\sim$

$$\left(\frac{\text{lb. fuel}}{\text{mile}}\right)_{\min} \sim \left(\frac{\text{lb. fuel}}{\text{hour}} \times \frac{1}{\text{miles/hour}}\right)_{\min} \sim \left(\frac{\text{lb. fuel}}{\text{b.hp.} \times \text{hour}} \times \frac{\text{b.hp.}}{V}\right)_{\min}$$

$$\sim \left(c \times \frac{P\eta}{V} \times \frac{1}{\eta}\right)_{\min} \sim \left(\frac{c}{\eta} \times \frac{TV}{V}\right)_{\min} \sim \left(\frac{c}{\eta} \times D\right)_{\min} \sim D_{\min} \sim (L/D)_{\max}.$$

Hence, if $c/\eta = $ constant, maximum range would be obtained by always flying at the attitude for maximum L/D.

(2) Maximum endurance $\sim \left(\frac{\text{fuel}}{\text{hr.}}\right)_{\min} \sim \left(\frac{\text{fuel}}{\text{b.hp.} \times \text{hr.}} \times \text{b.hp.}\right)_{\min}$

$\sim \left(c \times \frac{\text{t.hp.}}{\eta}\right)_{\min} \sim \left(\frac{c}{\eta} \times P_r\right)_{\min} \sim P_{r\min}.$ Hence, for constant c/η,

maximum endurance would be obtained by always flying at the attitude for minimum power required.

Let us consider the range problem in a little more detail following the procedure given by Breguet many years ago. The fundamental relation for both range and endurance is

$$\frac{dW}{dt} = - \text{ b.hp.} \times c$$

where $- dW = $ decrease in weight of the airplane due to burning fuel in the time interval dt.
Therefore

$$\frac{dW}{dt} = - \frac{\text{t.hp.}}{\eta} \times c = - \frac{c}{\eta} VD$$

$$dt = - \frac{\eta}{c} \frac{dW}{VD}$$

Now the rotal range R, flown in a time T is given by

$$R = \int_0^T V dt = - \int_{W_0}^{W_1} \frac{\eta}{c} \frac{dW}{D}$$

where

$W_0 = $ initial gross weight

$W_1 = $ final gross weight $= W_0 - W_F$

$W_F = $ lb. of fuel consumed

We may write

$$D = \frac{W}{L/D}$$

so that

$$R = \int_{W_1}^{W_0} \frac{\eta}{c} \cdot \frac{L}{D} \cdot \frac{dW}{W}$$

Breguet now assumes that, even though the flight is not carried out at maximum L/D, it will be conducted at constant angle of attack so that L/D is constant. Then since η/c is also assumed to be constant

$$R = \frac{\eta}{c} \frac{L}{D} \int_{W_1}^{W_0} \frac{dW}{W} = \frac{\eta}{c} \frac{L}{D} \log_e \frac{W_0}{W_1}$$

In engineering units

$$R \text{ (miles)} = 863.5 \frac{\eta}{c} \frac{L}{D} \log_{10} \frac{W_0}{W_1} \qquad \text{Breguet's formula} \qquad [3\cdot60]$$

The difficulty in using this relatively simple formula lies in the choice which must be made as to the best average values to use for η/c and L/D. From a study of many airplanes Diehl has developed a modified range formula which uses initial instead of average values. Diehl's formula is

$$R \text{ (miles)} = 625 \frac{\eta_0}{c_0} \left(\frac{L}{D}\right)_{\text{max}} \left[1 - \left(\frac{W_1}{W_0}\right)^{0.6}\right]$$

$$= K_R \frac{\eta_0}{c_0} \left(\frac{L}{D}\right)_{\text{max}} \qquad\qquad [3\cdot61]$$

where η_0 and c_0 correspond to the initial conditions. K_R is given by Diehl * as a function of W_F/W_0.

Analogous analyses and formulas have been developed by Breguet and Diehl for the problem of endurance, but these will not be considered here in view of the little practical application of this problem. For extremely accurate calculations of either range or endurance a lengthy tabular and graphical procedure must be followed.

(b) *Take-off.* The problem which must be solved in connection with take-off is the distance, S_0, required to accelerate from rest to a take-off speed, V, at which the airplane leaves the ground. This take-off speed is usually taken as 10 or 15 per cent greater than the stalling speed in order that there may be adequate control during the initial climb. A number of empirical and semi-theoretical expressions for S_0 have been given, the most frequently used being that of Diehl*

$$S_0 = K_s \frac{V^2 W}{T_i} \qquad\qquad [3\cdot62]$$

* *Loc. cit.*

where T_i is the initial thrust in pounds and V is the take-off speed in feet per second. The general form of equation $3 \cdot 62$ can easily be shown to be correct by the following very simple analysis: If v = velocity during the take-off run, a = acceleration, and t = time since the start of the run, then

$$v = \frac{dS_0}{dt}, \quad a = \frac{dv}{dt}$$

Therefore

$$dt = \frac{dv}{a}$$

and

$$S_0 = \int v \, dt = \int \frac{v \, dv}{a}$$

If now we consider the simplest case where a = constant, then

$$S_0 = \frac{1}{a} \int v \, dv = \frac{V^2}{2a}$$

But if T_e = effective thrust

$$T_e = \frac{W}{g} a = \text{constant}$$

Therefore

$$a = \frac{T_e g}{W}$$

Therefore

$$S_0 = \frac{1}{2g} \frac{V^2 W}{T_e} = 0.0155 \frac{V^2 W}{T_e}$$

Diehl's coefficient K_s replaces the factor 0.0155 and takes into account the variation in effective thrust. This can be expressed in terms of the ratio of the final net thrust to the initial thrust and normally has a value of the order of 0.025. The exact value depends on the engine-propeller combination characteristics and on the ground friction as is shown in detail in Diehl's discussion.

(c) *Cruising charts.* With the development of modern high performance airplanes and of intensely competitive commercial airlines it became essential to control the operation of airliners with great accuracy and economy. In 1934 E. T. Allen and W. B. Oswald * presented a series of fundamental papers on the subject of controlled operation of airplanes, which effectively revolutionized certain aspects of airline operations. One of the most important elements in the papers was the

---

* Cf. a series of articles by these authors which appeared in 1934 in *Aviation,* McGraw–Hill Book Co.

introduction of the cruising chart.   Allen and Oswald discussed many different types of these charts, of which only the simplest will be considered here: that giving brake horsepower for level flight as a function of true speed and altitude.   The following functional analysis furnishes the basis for such a chart.

We consider a given type of airplane equipped with constant-speed propellers and assume that a predetermined cruising revolutions per minute is used at all altitudes and velocities.   From equation $3 \cdot 26$ we have

$$P_r = F_1 (\sigma, V)$$

since l_s and l_p have definite numerical values for a given type of plane. Turning to power available, d and N are supposed given so that J is determined from V and hence using propeller charts

$$\eta = F_2 (V, \beta) \quad \text{and} \quad C_P = F_3 (V, \beta)$$

Now $P = C_P \rho n^3 d^5$, where $P =$ brake horsepower, so that

$$P = F_4 (V, \beta, \sigma)$$

Futhermore $P_a = P \eta$ and hence

$$P_a = F_5 (V, \beta, \sigma)$$

We are considering level flight so that $P_a = P_r$, and

$$F_1 (V, \sigma) = F_5 (V, \beta, \sigma)$$

which may be solved for β giving

$$\beta = F_6 (V, \sigma)$$

Substituting this in F_4 we obtain finally

$$P = F (V, \sigma) \qquad\qquad [3 \cdot 63]$$

which shows that for a given airplane whose propellers operate at a definite, constant revolutions per minute the brake horsepower is uniquely determined by true airspeed and density.   This relation could be represented by a diagram of the form shown in Fig. $3 \cdot 20$.   The constant-power curves can be determined by a flight test in which P, σ, and V are observed.   In general a manufacturer building a commercial transport for an airline will be required to furnish such a cruising chart with the first airplane of a series.

In actual practice the density scale is replaced by a density-altitude scale, and a pressure altitude-density altitude conversion chart is attached to the left of the diagram so that the final chart looks like

Fig. 3·21. The example indicated on the figure illustrates the solution to the following pilots' problem: The plane is at 8,000-foot pressure altitude, the air temperature is 60° F., what power must the engines be run at to maintain a speed of 190 m.p.h.? The answer is 70 per cent of

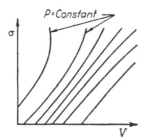

Fig. 3·20.   Power-density-velocity relation for level flight.

full rated power.   It is obvious that such a cruising chart will be of great assistance to a pilot making scheduled flights.   It also gives a very complete picture of the level-flight performance characteristics of a new airplane.

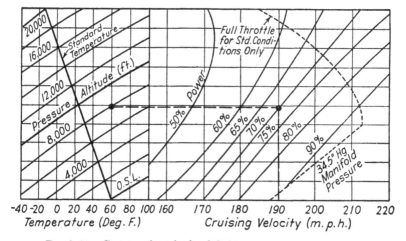

Fig. 3·21.   Cruising chart for level flight, constant propeller r.p.m.

(d) *Performance flight testing.*   This problem which is actually very complex, can here be discussed only briefly.   The first and most fundamental step in a flight testing schedule is to obtain reliable calibrations of *all* instruments to be used.   The next is a calibration of the air-speed indicator which is usually accomplished by flying over a measured speed course at very low altitude and comparing true ground speed with indicated air speed. Next the level-flight characteristics are determined in essentially the manner outlined above for the preparation of a cruising chart. Speed for best climb is then investigated by observing rate of climb at several values of V in the neighborhood of the anticipated V_c. This is repeated at a number of altitudes. Finally a climb is carried out from sea level to nearly the ceiling while the velocity is held at the speed for best climb as previously determined. During this climb rate of climb is observed as a function of altitude. The results of this climb test must be corrected to standard conditions, using a procedure which is too complicated to be reproduced here. Many additional performance items are usually investigated in addition to the above elements of the classical performance problem. Examples are: partial engine performance, fuel consumption tests, take-off run, landing space required in landing over a 50-foot obstacle.

CHAPTER 4

LONGITUDINAL STABILITY AND CONTROL

4-1. General Discussion of Stability; Fundamental Equations for Static Longitudinal Stability

In the remainder of this work the airplane will be considered as a rigid body having certain inertia characteristics and acted upon by external forces and moments which arise from the aerodynamic reactions between the airplane and the air through which it moves. The control surfaces may be displaced in order to obtain these forces and moments but no other distortion of the airplane is considered. The attainment of certain equilibrium attitudes or types of motion and the stability of these motions when the equilibrium is disturbed through changes in the aerodynamic forces and moments are here of primary importance. In mechanics discussion of the stability of a given equilibrium state includes consideration of the system in question when it is slightly displaced from its equilibrium condition, and investigation of how the system returns to or continues to move away from the equilibrium state. In aerodynamics such a complete discussion of the motion after a disturbance is listed under the title of "dynamic" stability. The term "static" stability is usually used to denote the much more restricted problem of the initial tendency of the system to return or move away from the equilibrium condition after a disturbance. In a static stability discussion it follows that the complete motion is not discussed at all, and when a system is said to be statically stable it means only that the system initially tends to return to the equilibrium condition when disturbed. It may, however, oscillate about the equilibrium condition without ever remaining in it. In such a case the system although statically stable may be dynamically unstable.

One further item of notation is important. Airplane motions may be divided into two categories, the first, called "longitudinal," in which the plane of symmetry remains in its original position, and the second, called "lateral," in which the plane of symmetry moves out of its original plane. The first is primarily connected with pitching motions, the second with rolling and yawing. It can be shown that for small disturbances the two motions are entirely independent, so that they may be discussed quite independently of one another. In this chapter the study will be limited to longitudinal motions which occur when an airplane is slightly disturbed from steady rectilinear flight. Only static stability will be considered until the final section.

Of the many different procedures proposed for the discussion of static longitudinal stability, only one which has been very widely adopted for quantitative analysis in this country will be considered here. This involves the plotting and study of curves of pitching moment coefficient about the center of gravity vs. C_L (or at times α). Two representative curves of this type are shown in Fig. 4·1. The condition for equilibrium, or trim as it is called, is obviously $C_M = 0$. Consider an airplane, represented by the solid curve, flying at the C_L for trim so

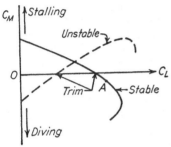

FIG. 4·1. Representative pitching moment curves.

that the resultant moment about the center of gravity is zero, and assume that the plane flies into an upward gust which increases the angle of attack. Then C_L is increased above the trim value. The moment coefficient now corresponds to the point A on the curve, so that a negative or diving moment acts on the airplane. This tends to nose the plane down, decrease the angle of attack, and return the airplane to its original trim attitude. Similarly a down gust decreases C_L which leads to a stalling moment, tending to increase the angle of attack and C_L towards their original equilibrium values. It is clear that a negative slope of the curve of C_M vs. C_L corresponds to static longitudinal stability. It is also easy to see that a positive slope as exemplified by the dotted curve corresponds to instability. The degree of stability or instability is obviously measured by the slope of the curve, and this can therefore serve as a convenient quantitative measure of static stability. We accordingly define

$$-\frac{dC_M}{dC_L} = \text{Static longitudinal stability} \qquad [4·1]$$

The two elements of a normal airplane which primarily determine its stability characteristics are the wing and horizontal tail surfaces. The study of stability will therefore be based on a skeleton airplane, composed of a center of gravity, a wing, and a horizontal tail, in gliding flight without thrust acting. The effects of propeller thrust, fuselage, nacelles, etc., will later be taken into account as correction factors.

Figure 4·2 shows such a skeleton airplane together with the aerodynamic forces acting on it. The most important items of notation are also indicated. It will be noted that the aerodynamic forces have been resolved into components in two ways: one relative to the reference axis (usually taken parallel to the fuselage axis or to the thrust axis) and the other referred to the flight path or relative wind direction. The intersection of the wing force vector with the wing chord line is usually called the center of pressure. By taking moments about the center of gravity and

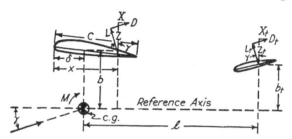

FIG. 4·2.   Forces and notation for skeleton airplane.

letting the subscript $(\ )_t$ denote "tail" while $(\ )_w$ denotes "wing," the following equations result:

$$M = M_w + M_t$$
$$M_w = - Z_w(x - \delta) + Xb$$
$$M_t = - Z_t l + X_t b_t$$

[4·2]

These equations are exact. In order to handle them certain approximations are made:

(a) The wing chord line is nearly parallel to the reference axis, so that x and δ may be measured parallel to either.
(b) $Z_w = L_w = L$ (total) $= L$
(c) $Z_t = L_t$
(d) $|\,X_t b_t\,| \ll |\,Z_t l\,|$
(e) l = constant independent of α. l is usually taken as the horizontal distance from center of gravity to elevator hinge line.

Assumption (b) states that the wing normal force is numerically approximately equal to the wing lift, which is an experimentally justified approximation. It also states that the tail (and other parts of the airplane) make a negligible contribution to the total lift. For extremely refined analyses it may be necessary to consider the contribution of the tail surfaces to the total lift, but this is usually not necessary. (d) is based on the experimental fact that the contribution of tangential force to the tail moment is negligible for all normal airplane arrangements. (e) rests on the fact that the motion of the tail surface center of pressure is a negligibly small fraction of the total tail length l.

Introducing these assumptions equations 4·2 become

$$\left.\begin{aligned} M_w &= -L(x - \delta) + Xb \\ M_t &= -L_t l \end{aligned}\right\} \qquad [4\cdot3]$$

In order to obtain dimensionless coefficients it is necessary to divide both equations by $q \times$ (an area) $\times$ (a length). The convention has universally been adopted of choosing the "area" as the wing area, S, and the "length" as the mean wing chord, c. In accordance with the usual conventions the following coefficients are defined

$$\left.\begin{aligned} C_M &= \frac{M}{qSc} & \qquad C_L &= \frac{L}{qS} \\[2mm] C_{M_w} &= \frac{M_w}{qSc} & \qquad C_X &= \frac{X}{qS} \\[2mm] C_{M_t} &= \frac{M_t}{qSc} & & \end{aligned}\right\} \qquad [4\cdot4]$$

With regard to L_t a different convention is more convenient. The tail-lift coefficient is naturally defined in terms of the tail-surface area, S_t, and the average dynamic pressure at the tail, q_t, which, because of interference effects, is often quite different from the free-stream dynamic pressure, $q = \frac{1}{2}\rho U^2$. Therefore

$$C_{L_t} = \frac{L_t}{q_t S_t} \qquad [4\cdot5]$$

Dividing equations 4·3 by qSc we now obtain

$$\left.\begin{aligned} C_M &= C_{M_w} + C_{M_t} \\[2mm] C_{M_w} &= -\left(\frac{x}{c} - \frac{\delta}{c}\right) C_L + \frac{b}{c} C_X \\[2mm] C_{M_t} &= -\frac{q_t}{q}\frac{l}{c}\frac{S_t}{S} C_{L_t} \end{aligned}\right\} \qquad [4\cdot6]$$

These various terms will be treated in greater detail in the following sections.

4–2. Effect of Longitudinal Shift in Center of Gravity; Wing Moments with Center of Gravity on the Chord Line

Probably the most important parameter in connection with longitudinal stability is the longitudinal location of the center of gravity relative to the wing, i.e., δ, and one of the problems most frequently confronting the aerodynamicist is the question of the effect on stability of a small fore-or-aft shift in the center of gravity location. In general the shifts in question will be so small that their effect in changing the tail length l is negligible. This means that the tail moment is essentially unaffected by such a change. Furthermore since the change is a longitudinal one, b is not altered and the moment contribution bC_X also remains unaffected. If the moment coefficients relative to two center of gravity positions, δ_1 and δ_2, are denoted by C_{M_1} and C_{M_2}, respectively, and if the two moments are compared at the same C_L, it follows that

$$C_{M_1} - C_{M_2} = -\left(\frac{x}{c} - \frac{\delta_1}{c}\right)C_L + \left(\frac{x}{c} - \frac{\delta_2}{c}\right)C_L = \left(\frac{\delta_1}{c} - \frac{\delta_2}{c}\right)C_L$$

Hence

$$C_{M_2} = C_{M_1} + \left(\frac{\delta_2}{c} - \frac{\delta_1}{c}\right)C_L \qquad [4\cdot7]$$

In multiplying by -1 and differentiating with respect to C_L to obtain the slopes of the moment curves, i.e., the stabilities, we obtain

$$\left(-\frac{dC_M}{dC_L}\right)_2 = \left(-\frac{dC_M}{dC_L}\right)_1 - \left(\frac{\delta_2}{c} - \frac{\delta_1}{c}\right) \qquad [4\cdot8]$$

Hence moving the center of gravity aft ($\delta_2 > \delta_1$) decreases the stability by an amount equal to the fraction of the chord length moved. Similarly moving the center of gravity forward increases the stability by a corresponding amount. The reason for the extreme importance of longitudinal center of gravity position in connection with stability questions is, therefore, clear.

The above considerations lead very easily to a determination of the character of the wing-moment curve when the center of gravity is on the chord line. Equation $4\cdot7$ gives the relation between the wing moments about any two points located distances δ_1 and δ_2 aft of the chord-line leading edge, i.e.,

$$C_{M_{w_2}} = C_{M_{w_1}} + \left(\frac{\delta_2}{c} - \frac{\delta_1}{c}\right)C_L$$

Let δ_1 be the distance of the aerodynamic center aft of the leading edge and let δ_2 give the center of gravity location. Following the usual convention we write

$\delta_1 = h =$ distance of aerodynamic center from leading edge.

$C_{M_1} = C_{M_0} =$ moment coefficient about the aerodynamic center.

$\delta_2 = \delta$ corresponding to the actual center of gravity location.

$C_{M_{w_2}} = C_{M_w} =$ moment coefficient about the center of gravity.

Then

$$C_{M_w} = C_{M_0} + \left(\frac{\delta}{c} - \frac{h}{c}\right) C_L \qquad \text{Wing moment about center of gravity on chord line} \qquad [4 \cdot 9]$$

Earlier, the discussion of airfoil characteristics showed that the moment coefficient about the aerodynamic center is approximately constant and the aerodynamic center is located near the 24 per cent point for normal wings. Hence

$$\left. \begin{array}{c} \dfrac{h}{c} \doteq 0.24 \\[2em] C_{M_0} \doteq \text{Constant} \end{array} \right\} \qquad [4 \cdot 10]$$

If $C_{M_0} =$ constant then from equation $4 \cdot 9$

$$\left(-\frac{dC_{M_w}}{dC_L} \right) = -\left(\frac{\delta}{c} - \frac{h}{c}\right) \doteq 0.24 - \frac{\delta}{c} \qquad [4 \cdot 11]$$

Therefore, an airplane having only a wing must have the center of gravity ahead of the aerodynamic center if it is to be stable. Figures $4 \cdot 3$ (a) and $4 \cdot 3$ (b) show typical wing-moment curves for airplanes having

(a) a normal cambered airfoil section, $C_{M_0} < 0$.

(b) an airfoil with reflexed mean camber line, $C_{M_0} > 0$.

For the normal airfoil it is impossible to obtain both trim and stability as appears from Fig. $4 \cdot 3$ (a). Both can be achieved, however, if a reflexed airfoil, having a positive C_{M_0} is used and the center of gravity is located ahead of the aerodynamic center. This is exactly the combination which has been successfully employed on the various "tailless" airplanes which have been constructed. In fact it is the arrangement which must be used on such airplanes.

Before leaving this section it should be explicitly pointed out that the validity of equation $4 \cdot 9$ does not depend at all on the assumption that

C_{M_0} is a constant. The equation is still true if C_{M_0} is itself a function of C_L. Here, however, equation 4·11 does not hold, since, in the differentiation which led to it, dC_{M_0}/dC_L was set equal to zero.

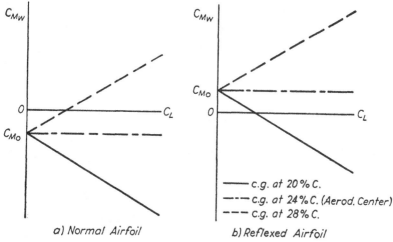

a) Normal Airfoil   b) Reflexed Airfoil

Fig. 4·3.   Representative wing-moment curves.

4–3. Horizontal Tail Moments

Equation 4·6 gives the fundamental expression for tail-moment coefficient

$$C_{M_t} = -\frac{q_t}{q}\frac{l}{c}\frac{S_t}{S}\,C_{L_t} \qquad\qquad [4\cdot12]$$

This is to be combined with the corresponding wing-moment coefficient which, as has been shown, is a function of C_L. In order to put C_{M_t} into

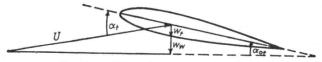

Fig. 4·4.   Velocity components at the tail.

a comparable form the dependence of C_{L_t} on C_L must be found. This is unfortunately complicated by the fact that the tail surfaces lie in the downwash produced by the wing. It is convenient to consider this point in some detail before proceeding to the analysis of C_{L_t} itself.

Figure 4·4 shows the various velocity components at a tail surface which, for simplicity, has been represented as a symmetrical section so that the chord line is also the zero lift line. w_t represents the downwash due to the trailing vortices leaving the tail itself and w_w is that due to the wing trailing vortices. The geometrical and effective angles of attack of the tail zero lift line are shown as α_t and α_{0_t}, respectively. Elliptic lift distribution is assumed over the tail surface so that in accordance with equation 1·28

$$\frac{w_t}{U} = \frac{C_{L_t}}{\pi Æ_t} \qquad [4·13]$$

For the wing downwash w_w at the tail, Fig. 4·5 shows schematically the theoretical and actual downwash distribution along a horizontal line in the plane of symmetry of a wing. The downwash at the lifting

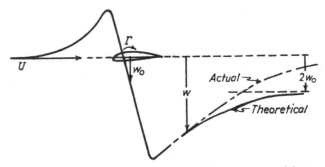

Fig. 4·5.   Downwash distribution before and behind a wing of finite span.

line is denoted by w_0, which, by assuming elliptic lift distribution, is given as

$$\frac{w_0}{U} = \frac{C_L}{\pi Æ} \qquad [4·14]$$

where $Æ = Æ_w$ aspect ratio of the wing. Close behind the wing the effect of the bound vortex adds to that of the trailing vortices to give $w \gg w_0$. As the distance downstream increases, the effect of the bound vortex decreases and rapidly approaches zero. However, that of the trailing vortices increases, since for a point behind the wing these vortices not only extend downstream to infinity but also upstream to the lifting line. Finally, for a point far downstream the trailing vortices appear to extend to infinity in both the downstream and upstream directions, so that the downwash produced by them has twice the value of that at the wing where the inducing vortices extend to infinity in only one direction. The theoretical downwash distribution accordingly has the characteristic shown by the solid curve of Fig. 4·5, and approaches $2w_0$ asymptotically at infinite distance downstream. Because of the action of viscosity and other secondary factors the trailing vortices tend to die out as the distance from the wing increases so that the actual downwash far downstream is smaller than the theoretical value, approaching zero rather than $2w_0$ asymptotically. It is clear that the value w_w at the tail surfaces depends on the relative position of wing and tail, and many theoretical and experimental investigations on this point have been carried out.* A very simple assumption has been used for the past 12 years at the GALCIT with considerable success. This is that the actual downwash at the tail equals the theoretical value at infinity, i.e.,

$$\frac{w_w}{U} = \frac{2w_0}{U} = 2\frac{C_L}{\pi R} \qquad [4 \cdot 15]$$

The present analysis is based on this assumption which appears to be entirely satisfactory in most cases, at least as a first approximation.

One further item of nota-
tion must be discussed before proceeding to the actual
analysis of the relation
between C_{L_t} and C_L. Figure
4·6 shows geometrical an-
gles of attack of the zero
lift lines for wing and tail.

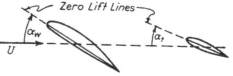

Fig. 4·6. Geometric angles of attack of wing and tail.

The difference between the two is called the aerodynamic decalage and is defined by

$$\alpha_d = \alpha_w - \alpha_t \qquad \text{Aerodynamic decalage} \qquad [4 \cdot 16]$$

The decalage for an airplane can most conveniently be determined by the following procedure: Consider the airplane as a whole rotated until the tail surface would be at zero lift neglecting all downwash effects, i.e., $\alpha_t = 0$. Then the geometrical angle of attack of the wing above zero lift is the decalage α_d. α_d may be varied on an actual airplane by changing the stabilizer setting, or by deflecting the elevators and so changing the zero lift line of the tail surfaces. It is also altered when wing flaps are deflected changing the zero lift line of the wings. On conventional airplanes the decalage is always positive, i.e., the wing is at a larger angle of attack than the tail.

In carrying out the analysis it is assumed that both wing and tail have the same infinite aspect ratio lift curve slope, a_0. Then

$$C_{L_t} = a_0\alpha_{0_t} = a_0\left(\alpha_t - \frac{w_t + w_w}{U}\right) = a_0\left(\alpha_t - \frac{w_t}{U} - \frac{w_w}{U}\right)$$

* Cf. Abe Silverstein and S. Katzoff, "Design Charts for Predicting Downwash Angles and Wake Characteristics Behind Plain and Flapped Wings," *N.A.C.A. Tech. Rep.* 648 (1939).

from the geometry of Fig. 4·4. Now introducing equations 4·13 and
4·15

$$C_{L_t} = a_0\alpha_t - \frac{a_0}{\pi R_t}C_{L_t} - \frac{2a_0}{\pi R}C_L$$

By collecting the terms in C_{L_t} and by replacing α_t by $(\alpha_w - \alpha_d)$ in
accordance with equation 4·16

$$\left(1 + \frac{a_0}{\pi R_t}\right)C_{L_t} = a_0\alpha_w - 2\frac{a_0}{\pi R}C_L - a_0\alpha_d$$

Now, since it is assumed that $C_L = C_{L_w}$ and that the lift distribution
over the wing is elliptical

$$C_L = a_0\left(\alpha_w - \frac{w_0}{U}\right) = a_0\alpha_w - \frac{a_0}{\pi R}C_L$$

Therefore

$$a_0\alpha_w = \left(1 + \frac{a_0}{\pi R}\right)C_L$$

so that

$$\left(1 + \frac{a_0}{\pi R_t}\right)C_{L_t} = \left(1 - \frac{a_0}{\pi R}\right)C_L - a_0\alpha_d$$

and finally

$$C_{L_t} = \frac{1 - (a_0/\pi R)}{1 + (a_0/\pi R_t)}C_L - \frac{a_0}{1 + (a_0/\pi R_t)}\alpha_d \qquad [4·17]$$

By introducing this into equation 4·12 the tail moment coefficient is

$$C_{M_t} = -\frac{q_t}{q}\frac{l}{c}\frac{S_t}{S}\left(\frac{1 - (c_0/\pi R)}{1 + (a_0/\pi R_t)}C_L - \frac{a_0}{1 + (a_0/\pi R_t)}\alpha_d\right) \qquad [4·18]$$

In normal airplanes the tail surfaces lie downstream of the wing and
fuselage and also of the nacelles when such are present. Accordingly
the air striking the tail in gliding flight has a velocity which is considerably lower than that corresponding to the free stream. This means
that $q_t/q < 1$. Furthermore, the lift distribution over wing and tail is
not elliptical as was assumed in deducing equation 4·18, and the downwash conditions are more complicated than was assumed. All these
elements are taken into account by introducing an empirical "tail efficiency factor" η_t into equation 4·18 in place of q_t/q. The final expression is accordingly *

$$C_{M_t} = -\eta_t\frac{lS_t}{cS}\left(\frac{1 - (a_0/\pi R)}{1 + (a_0/\pi R_t)}C_L - \frac{a_0}{1 + (a_0/\pi R_t)}\alpha_d\right) \qquad [4·19]$$

* It should be pointed out that throughout the stability discussion consistent
units must be used for angles of attack and lift curve slopes. In particular if $a_0 = 5.7$,
the usual assumption, then α_d must be measured in radians and not in degrees.

In this expression S_t is taken as the total horizontal tail area including any area covered by the fuselage, and when no data for a specific airplane are available η_t may be chosen from the following table of empirical average values as determined over a period of years in the GALCIT wind tunnel.

AVERAGE EMPIRICAL VALUES OF η_t

Airplane Type	Usual Range of η_t	Mean η_t
Low wing landplanes and flying boats; single vertical tail	0.58–0.70	0.62
Low wing landplanes and flying boats; twin vertical tail	0.65–0.75	0.70
High wing monoplanes and biplanes; single vertical tail	. . . .	0.80

A study of equation 4·19 shows that for a given airplane everything on the right is constant except α_d and C_L. α_d may be varied by changing the tail surface setting but for a given setting is also constant. Hence the curves of C_{M_t} vs. C_L for a series of fixed tail settings (elevator angles) form a family of parallel straight lines with negative slope, cf. Fig. 4·7. It should be noted that negative elevator angles (denoted by e) correspond to trailing edge raised and to an increase in α_d. Positive e corresponds to trailing edge lowered and to a decrease in α_d. The stabilizing effect of the tail is not under the control of the pilot, but by deflecting elevators he can alter the trim conditions. Although the primary factor determining the stabilizing effect of

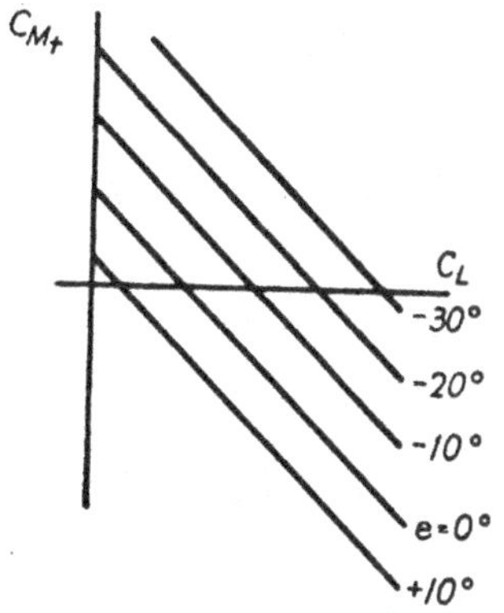

FIG. 4·7.   Tail moment curves for various elevator deflections.

the tail is the ratio lS_t/cS, i.e., the tail area and tail length, both wing and tail aspect ratio are also of importance, the stability being increased by an increase in the value of either one.

4–4. Moments for Complete Skeleton Airplane; Effects of Fuselage, Nacelles, Vertical Center of Gravity Location, and Propeller Thrust; Elevator Effect

The expressions for wing and tail moments have been found and can now be combined to furnish the total pitching moment for the skeleton airplane whose center of gravity lies on the wing chord line. Combining equations 4·6, 4·9, and 4·19 we have

$$C_M = \left[\left(\frac{\delta}{c} - \frac{h}{c}\right) - \eta_t \frac{lS_t}{cS}\left(\frac{1 - (a_0/\pi \mathcal{R})}{1 + (a_0/\pi \mathcal{R}_t)}\right)\right]C_L$$
$$+ \left[C_{M_0} + \eta_t \frac{lS_t}{cS}\frac{a_0}{1 + (a_0/\pi \mathcal{R}_t)}\alpha_d\right] \qquad [4·20]$$

If C_{M_0} is constant, as is usually the case, this represents a straight line for any given tail setting. Different tail settings correspond to different straight lines, but all are parallel to one another and all correspond to the same stability

$$-\frac{dC_M}{dC_L} = \eta_t \frac{lS_t}{cS}\left(\frac{1 - (a_0/\pi \mathcal{R})}{1 + (a_0/\pi \mathcal{R}_t)}\right) - \left(\frac{\delta}{c} - \frac{h}{c}\right) \qquad [4·21]$$

It is clear that the stability is a characteristic which is built into the airplane and cannot normally be varied in flight except by shifting weight fore or aft to change the center of gravity location. The designer's problem is to lay out his airplane so that it will have an adequate value for the stability with the furthest aft center of gravity position, and then to furnish enough tail surface effect to make trim possible at all C_L's from that corresponding to somewhat more than maximum level-flight velocity up to $C_{L\max}$. The stability portion of this problem is usually the more important and must be treated with the more accuracy.

The secondary effects which have hitherto been neglected may have an influence on both parts of the problem, i.e., on the two bracketed terms of equation 4·20. However, their effect on stability is more consistent and more significant than their influence on trim, which latter influence is usually quite small. Hence for an approximate treatment they are frequently considered as having no effect on trim and their effect on stability is usually expressed as the increment in dC_M/dC_L which is added when they are superimposed on the skeleton airplane. Both fuselage and nacelles have almost always destabilizing effects.

As a result of many tests at the GALCIT these may be expressed by the following average values when specific data are not available.

$$\left.\begin{array}{l} \text{Fuselage } \Delta \left(\dfrac{dC_M}{dC_L}\right) \doteq 0.02 \text{ to } 0.04; \text{ average } 0.03 \\[3ex] \text{Nacelles } \Delta \left(\dfrac{dC_M}{dC_L}\right) \doteq 0.025 \text{ to } 0.045; \text{ average } 0.035 \end{array}\right\} \qquad [4\cdot22]$$

The effect of vertical displacement of the center of gravity from the chord line is not so simple. As a rough first approximation, based on a study of the term $(b/c)C_X$ in equation $4\cdot6$, it has been found possible to give the general rule

$$\text{Vertical c.g. location } \Delta \left(\frac{dC_M}{dC_L}\right) \doteq -\frac{1}{10}\frac{b}{c} \qquad [4\cdot23]$$

This means that lowering the center of gravity (b positive) increases stability while raising the center of gravity (b negative) decreases the stability. Furthermore vertical center of gravity displacement has roughly one-tenth the effect of horizontal displacement on airplane stability.

All the above effects can readily be included in the moment equation $4\cdot20$ by replacing

$$\left.\begin{array}{l} \dfrac{\delta}{c} \rightarrow \dfrac{\delta_{\text{eff}}}{c} \text{ in equations } 4\cdot20 \text{ and } 4\cdot21 \\[3ex] \dfrac{\delta_{\text{eff}}}{c} = \dfrac{\delta_{\text{true}}}{c} - \dfrac{1}{10}\dfrac{b}{c} + \Delta\left(\dfrac{dC_M}{dC_L}\right)_{\text{fuselage}} + \Delta\left(\dfrac{dC_M}{dC_L}\right)_{\text{nacelles}} \end{array}\right\} \qquad [4\cdot24]$$

δ_{eff} is the effective horizontal center of gravity location for a skeleton airplane having approximately the same moment characteristics as the corresponding complete airplane with center of gravity location given by δ_{true}.

Before leaving these effects a warning in connection with vertical center of gravity location should be emphasized. In addition to changing the average stability as discussed above a center of gravity position off the chord line often introduces curvature into the moment curve of the type illustrated in Fig. $4\cdot8$. This effect may be particularly serious in the case of low-wing monoplanes whose center of gravity is above the chord line, since in such cases it takes the form of a rapid decrease in stability near the stall. Unfortunately the effect differs considerably from plane to plane and no satisfactory general method has yet been found of taking it into account. Several early low-wing monoplanes experienced difficulty, because this effect of high center of gravity location was not at first known or expected.

Any study of the effects of propeller thrust or slipstream reveals a very confused and unsatisfactory state of affairs. In most cases the running of propellers has a destabilizing effect which increases with the propeller thrust. In some airplanes this effect may be very large,

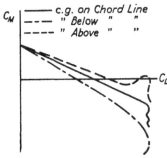

amounting to as much as $\Delta(dC_M/dC_L)$ $= 0.20$ near the stall. For other airplanes it is very much smaller. No simple, general rules have yet been discovered which will permit the designer to predict accurately the effect on a new design, and it has accordingly become increasingly common to install model propellers driven by tiny but powerful electric motors in wind-tunnel models. The true power conditions are then duplicated and the effect of power on stability determined before the full-size airplane is built.

Fig. 4·8. Effect of vertical c.g. location on moment-curve curvature.

The practice which has been recommended for some time at the GALCIT, when such power-on data are not available, is to require the power-off stability for the rearmost center of gravity position to have approximately the value

$$\left(-\frac{dC_M}{dC_L} \right) = 0.10 \text{ to } 0.15 \qquad [4\cdot25]$$

This has led to satisfactory results with some few exceptions. With the present state of knowledge it appears to be about as satisfactory a general rule as can be formulated.

Two remarks dealing with elevator effects must be made in concluding this section. The first is concerned with the effect of elevator displacement on aerodynamic decalage, α_d. Let

$$\alpha_d = \alpha_{d_0} + \Delta\alpha_d \qquad [4\cdot26]$$

where

α_{d_0} = value of α_d with elevator neutral ($e = 0$)

$\Delta\alpha_d$ = change in α_d due to elevator deflection

Then theory and experiment agree in showing that

$$\Delta\alpha_d = -\,Ke \qquad [4\cdot27]$$

where

e = elevator deflection, taken positive when trailing edge is lowered

K = a function of the ratio of elevator to total tail chord

K is given in Fig. 4·9 which is taken from Wood,* and corresponds to empirical modifications to a theoretical relation originally given by Glauert.†

The second remark is that estimates of the stabilizing effect of a tail made by the above procedure will usually be optimistic if the elevator is

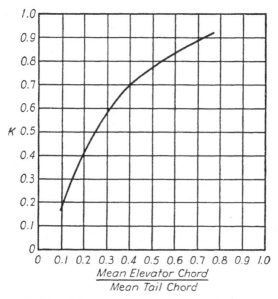

FIG. 4·9.   Elevator effectiveness factor.

allowed to float freely, as in hands-off flight, instead of being held at a fixed angle as was assumed in the analysis.

4–5. Dynamic Longitudinal Stability

When it is necessary to discuss the true stability of an airplane, i.e., the complete nature of its response to a disturbance from an equilibrium condition, the static stability characteristics are inadequate. For example, consider an airplane in steady horizontal flight with the refer-

* K. D. Wood, *loc. cit.*, Fig. 273.

† The remark contained in the footnote associated with equation 4·19 is applicable to $\Delta\alpha_d$ as well as to α_t itself.

ence axis at an angle θ_0 to the horizontal. We ask for the true stability of the pitching motion following a displacement of this angle θ from its equilibrium value θ_0, as the result of a gust. The answer to the problem is most conveniently given in the form of a graph of θ against time t. Figure 4·10 shows typical curves corresponding to the four types of motion which can occur. These types of motion may be calculated theoretically on the assumption that the displacements from the equilibrium position, $\theta - \theta_0$, are small.

Of the motions represented in the figure, two, numbered (1) and (2), are stable in that as $t \to \infty$, $\theta \to \theta_0$, while two, (3) and (4), are unstable.

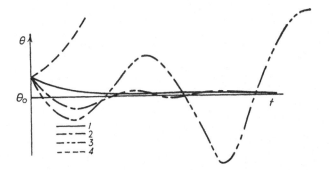

FIG. 4·10.   Typical airplane pitching motions.

However (1), (2), and (3) all satisfy the requirement for static stability in that the initial tendency is toward, rather than away from, the equilibrium condition. The four types of motion have been given the following descriptive designations:

(1) Simple subsidence, statically stable, dynamically stable.

(2) Damped oscillation, statically stable, dynamically stable.

(3) Divergent oscillation, statically stable, dynamically unstable.

(4) Divergence, statically unstable, dynamically unstable.

The actual longitudinal motion of an airplane as observed in flight in general falls into category (2), (3), or (4). Case (4) corresponds to static instability and has already been considered. Motions of the type (2) and (3) are characterized by two parameters: the period or time required for one complete oscillation, and the damping factor which determines the rate at which the amplitude of successive oscillations decreases, or increases (if the damping factor is negative so that the motion is unstable). For normal airplanes the longitudinal period is of the order of

20 to 60 seconds which is of such length as to be very noticeable. The damping factor, whether positive or negative, is very small so that the oscillations are slowly damped if the motion is stable and increase only slowly if the motion is unstable. Hence even if the airplane is unstable the motion is in no sense dangerous or difficult to control.

For a considerable period of time all airplanes licensed for transport operation in the United States were required to demonstrate that longitudinal oscillations were damped under all normal flight conditions. As a result most large aircraft manufacturing companies carried out elaborate dynamic stability calculations. Recently the feeling has been growing that the damping of these slow longitudinal oscillations is relatively unimportant, and the licensing requirements in this connection have been relaxed. The significance of longitudinal dynamic stability studies, is therefore, now reduced, so that it is not necessary to consider the subject in any further detail in this introductory work.*

---

* An excellent discussion with numerous design charts permitting a rapid estimate of the dynamic longitudinal stability characteristics of an airplane has been given by Charles H. Zimmerman, "An Analysis of Longitudinal Stability in Power-Off Flight with Charts for use in Design, *N.A.C.A. Tech. Rep.* 521 (1935).

CHAPTER 5

LATERAL STABILITY AND CONTROL

5-1. General Discussion of Lateral Stability.

To discuss the general stability of an airplane it is necessary to adopt a definite and rather elaborate system of notation since the motion is inherently complicated. Figure 5·1 shows the conventions which have

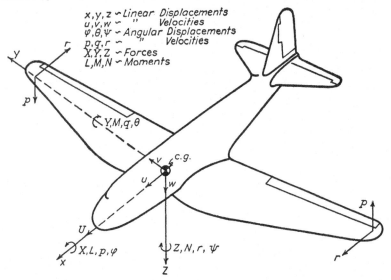

x,y,z – Linear Displacements
u,v,w – " Velocities
φ,θ,ψ – Angular Displacements
p,q,r – " Velocities
X,Y,Z – Forces
L,M,N – Moments

FIG. 5·1. Standard conventions for airplane stability discussion.

been officially adopted for the United States by the N.A.C.A. An airplane considered as a rigid body has six degrees of freedom so that its motion may be described by six variables, three giving mutually perpendicular velocity components of the center of gravity, and three giving rotations or angular velocities about three mutually perpendicular axes through the center of gravity. Two different sets of axes are used in both of which the y axis is parallel to the wing span. The first, called "wind axes," has its x axis in the direction of the undisturbed flight path so that the x and z components of the resultant aerodynamic force are

154

the drag and lift, respectively. The second, called "body axes," has its x axis parallel to the fuselage or thrust axis. For certain purposes wind axes are more convenient, while for others body axes are more useful. In our discussion here we shall generally use wind axes unless body axes are specifically mentioned.

The three velocity components of the center of gravity are denoted by u, v, w, and the three angular rotations by φ, θ, ψ as indicated in Fig. 5·1. From the figure it is clear that motions involving only u, w, and θ leave the plane of symmetry ($x - z$ plane) remaining in its own plane and hence are included in the field of longitudinal motions which we have already discussed. Motions involving v, φ, and ψ, or "sideslip," "roll," and "yaw" as they are called, are classified as lateral motions. The lateral motions of an airplane are inherently much more complicated than the longitudinal motions since sideslip, roll, and yaw are inextricably interconnected. Thus a pure sideslip, v, always produces a yaw, ψ, and a roll, φ. It has been shown that a pitching motion leads only to a pitching moment which affects only the pitching motion itself. However a sideslip velocity, v, not only produces a side force but also a rolling and a yawing moment, and these affect the rolling and yawing motion as well as the sideslip velocity itself. Only a single derivative, dC_M/dC_L or $dC_M/d\theta$ (since $C_L \sim \alpha \sim \theta$), is needed in the discussion of longitudinal stability, whereas in treating lateral motions a number of such derivatives must be employed.

Before considering these various derivatives it will be convenient to introduce a notation for coefficients which, although not that recommended by the N.A.C.A., has the advantage of avoiding certain confusions:

$$C_s = \frac{\text{Side force}}{qS} \qquad = C_Y \quad (\text{N.A.C.A.})$$

$$C_r = \frac{\text{Rolling moment}}{qsb} \qquad = C_L \quad (\text{N.A.C.A.}) \qquad [5\cdot1]$$

$$C_y = \frac{\text{Yawing moment}}{qsb} \qquad = C_N \quad (\text{N.A.C.A.})$$

It is not necessary to consider the effect of sideslip velocity separately since a sideslip velocity v has exactly the same aerodynamic effect as an angle of yaw $\psi = (-v)/U$, where U is the forward speed of the airplane (cf. Fig. 5·1). For the same reason the vertical velocity w was not considered in the longitudinal discussion since w would have the same effect as an increase in angle of attack (or of θ) of amount $\Delta\alpha = w/U$.

In connection with the moments pertinent to lateral stability it would, therefore, be expected that the following four derivatives would enter

$$\frac{\partial C_r}{\partial \varphi}, \quad \frac{\partial C_y}{\partial \psi}, \quad \frac{\partial C_r}{\partial \psi}, \quad \frac{\partial C_y}{\partial \varphi}$$

Actually two of these, $\partial C_r/\partial\varphi$ and $\partial C_y/\partial\varphi$, are zero as can easily be seen from Fig. 5·2 showing a front view of an airplane at an angle of roll, φ. The lift and weight forces (as well as the drag force not shown) produce no resultant moment, but only an unbalanced force, R. This side force would accelerate the plane sidewards and lead to a sideslip, i.e., to an

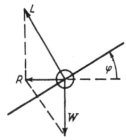

angle of yaw, ψ. This yaw in turn would produce moments, but this is a secondary phenomenon which depends on the yaw derivatives $\partial C_r/\partial\psi$ and $\partial C_y/\partial\psi$. The roll angle, φ, itself produces no moments at all.

Unfortunately this simplification is more than counterbalanced by a very serious complication. In view of the interrelationships between yaw and roll it is evident that the concept of static stability or "tendency to return to an equilibrium condition" has very little significance in connection with lateral motions.

Fig. 5·2.   Forces on a banked airplane.

Any discussion of lateral stability must actually deal with the complete dynamical motion of the airplane, and this means that not only angular positions but also angular velocities about the three axes are important. These are denoted, as shown in Fig. 5·1 by

$$p = \frac{d\varphi}{dt}, \quad q = \frac{d\theta}{dt}, \quad r = \frac{d\psi}{dt}$$

Each of these angular velocities produces moments, those which enter the lateral stability problem being given by the so-called rotary derivatives *

$$\frac{\partial C_r}{\partial p}, \quad \frac{\partial C_y}{\partial r}, \quad \frac{\partial C_r}{\partial r}, \quad \frac{\partial C_y}{\partial p}$$

* In the analysis of dynamic longitudinal stability the derivative $\partial C_M/\partial q$, i.e., the damping in pitch due to velocity of pitch, is of fundamental importance. Since dynamic longitudinal stability is much less important than (dynamic) lateral stability this longitudinal rotary derivative has not required discussion in the present work.

The six fundamental aerodynamic derivatives which control the lateral motion of an airplane may be collected as follows: *

$$
\left.\begin{array}{c} \dfrac{\partial C_y}{\partial \psi} \\[2mm] \dfrac{\partial C_r}{\partial \psi} \end{array}\right\} \text{Static derivatives}
$$

$$
\left.\begin{array}{c} \dfrac{\partial C_r}{\partial p} \\[2mm] \dfrac{\partial C_y}{\partial r} \end{array}\right\} \text{Damping derivatives}
$$

$$
\left.\begin{array}{c} \dfrac{\partial C_r}{\partial r} \\[2mm] \dfrac{\partial C_y}{\partial p} \end{array}\right\} \text{Cross derivatives}
$$

[5·2]

There are also derivatives connected with the side force, but they are of much less importance and can be omitted in the introductory treatment given here.

5-2. The Lateral Stability Derivatives

In this section the various lateral derivatives will be treated briefly in the order in which they are listed at the end of the preceding section. Considerations will be limited to conditions below the stall. It should be pointed out that the static derivatives can be determined experimentally in a normal wind tunnel whose rigging is so arranged that measurements can be made at angles of yaw. This is not true of the other four derivatives whose values must be estimated by calculations, or else obtained with great difficulty by using special devices which permit the forces and moments acting on a model to be measured instantaneously while the model is rotating or oscillating.

$\partial C_y/\partial \psi \sim$ *Directional stability $\sim$ yawing moment due to yaw.* This derivative, sometimes called the directional or weathercock stability,

* The use of r both as a subscript denoting roll and as a symbol for yawing velocity represents a disadvantage of the notation adopted. However the aeronautical engineer who is not a stability specialist will almost never deal with any except the static derivatives of 5·2, so that he will not need to use r as a symbol for yawing velocity. The notation here adopted avoids the confusion introduced through the use of L as a subscript signifying both "lift" and "rolling moment."

is of extreme importance. It is analogous to dC_M/dC_L in that it must have a negative value for stability. It is primarily determined by the product of the vertical tail surface area, S_v, and the tail length l_v. As a result of many tests at the GALCIT limits for the directional stability on satisfactory airplanes may be set as $0.0007 < (-\partial C_y/\partial \psi°) < 0.0015$.

$\partial C_r/\partial \psi \sim$ *Rolling moment due to yaw.* This derivative is primarily produced by dihedral and is, in fact, approximately proportional to dihedral. Figure 5·3 shows a front view of an airplane which has a positive yaw ($\psi > 0$, nose to the right). The relative wind therefore strikes the plane from the left. In view of the dihedral the angle of attack on the left wing is larger than that on the right so that the lift on the left is greater than that on the right as is indicated, and a positive rolling moment results. Hence $\partial C_r/\partial \psi$ is positive. The result is that

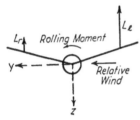

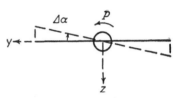

FIG. 5·3. Effect of dihedral combined with sideslip.

FIG. 5·4. Effect of rolling velocity on angle of attack distribution.

when such a plane drops a wing and sideslips toward the lower wing, the dihedral gives a rolling moment which tends to lift the low wing. Modern low-wing monoplanes are usually constructed with 5° to 8° dihedral in order that this derivative may have a satisfactory value.

$\partial C_r/\partial p \sim$ *Damping in roll.* This derivative gives the rolling moment produced by angular velocity of roll. It is always negative below the stall and hence represents a true damping, as is indicated in Fig. 5·4. This figure shows the distribution of angle of attack increment along the span produced by a positive velocity of roll, $p > 0$. The effective angle of attack of the down-moving wing is increased and that of the up-moving wing is decreased. Below the stall lift is proportional to α so that the lift on the right wing is increased and that on the left wing is decreased. The result is a powerful negative rolling moment which opposes the rolling motion and effectively damps out any purely rolling oscillations below the stall.

$\partial C_y/\partial r \sim$ *Damping in yaw.* The yawing moment due to velocity of yaw is also a true damping and hence is negative. The effect is produced both by the vertical surfaces and by the wings. The linear velocities produced by a positive angular velocity of yaw, $r > 0$, are shown schematically in the airplane top view of Fig. 5·5. The velocity of the vertical surface obviously gives a side force which leads to a negative yawing moment. When the wing is considered it is seen that the relative wind speed over the left wing is increased and that over the right wing is decreased by the rotation. Accordingly the drag on the left wing is increased and that on the right wing is decreased, since drag is proportional to (velocity)^2. This also gives a negative yawing moment which adds to that produced by the vertical surfaces.

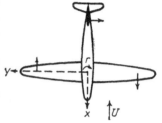

FIG. 5·5.   Effect of yawing velocity.

$\partial C_r/\partial r \sim$ *Rolling moment due to yawing velocity.* The diagram of Fig. 5·5 may again be utilized in discussing this derivative. The main effect here is due to the wings, the left wing having a higher lift and the right one a lower lift, since lift as well as drag is proportional to (velocity)^2. This unsymmetrical lift distribution gives a rolling moment tending to depress the right wing, i.e., a positive moment. Accordingly $\partial C_r/\partial r > 0$ which means that a plane which is yawing tends to bank in the direction of the turn (right wing down if the nose is turning to the right).

$\partial C_y/\partial p \sim$ *Yawing moment due to rolling velocity.* In connection with Fig. 5·4 it may be recalled that a positive roll gives an increase in lift on the right wing and a decrease on the left wing. The change in direction of the relative wind produced by the roll also results in a forward inclination of the lift vector on the right side and a rearward inclination on the left. This produces a yawing moment tending to move the right wing forward, i.e., a negative yawing moment. The unsymmetrical drag forces associated with the unsymmetrical lift distribution furnish an opposing or positive yawing moment. However the latter is always smaller than the moment produced by the inclined lift forces. Hence $\partial C_y/\partial p < 0$, i.e., in a rolling airplane the up-going wing is retarded and the down-going wing is accelerated forward.

All these derivatives can be investigated in much greater detail and formulas deduced which permit numerical values to be assigned with some accuracy.* However, it is believed that all the essential physical elements have been introduced in the above abbreviated survey.

* Cf. Fred E. Weick and Robert T. Jones, "The Effect of Lateral Controls in Producing Motion of an Airplane as Computed From Wind Tunnel Data," *N.A.C.A. Tech. Rep.* 570, Appendix I (1936).

5-3. Lateral Stability and Control Below the Stall

The dynamical analysis of the lateral motion of an airplane involves all the above derivatives as well as the moments of inertia about the lateral and vertical axes. The analysis is far too complex to enter into here,* but the results can be briefly summarized. It appears that three types of lateral instability can occur below the stall.

(a) *Directional Divergence.* Here the airplane executes a simple yawing divergence as if it were turning around to fly tail first. This instability only appears if $dC_y/d\psi$ is positive so that the plane has weathercock instability. The problem can therefore be considered as one of static stability and its solution requires only the furnishing of sufficient vertical tail surface to ensure weathercock stability.

(b) *Spiral Instability.* This is also a non-oscillatory or divergent motion, but is much less violent than the directional divergence. If a spirally unstable airplane, through the action of a gust or other disturbance, gets a small initial yaw, to the right for example, this is followed by a gentle bank to the right. This bank causes additional yaw which is followed by more bank and the process continues. The motion starts as a gentle spiral which, as time goes on, gets continuously tighter and steeper until finally, if the motion is not checked, a steep, high speed, spiral dive results. The motion develops so very gradually, however, that it is usually corrected unconsciously by the pilot who is often completely unaware that spiral instability exists for a spirally unstable airplane. The combination of high weathercock stability and small dihedral leads to spiral instability. Conversely increasing dihedral and decreasing directional stability reduces the tendency towards this type of instability.

(c) *Lateral Oscillations or "Dutch Roll."* In normal airplanes this oscillatory motion is never actually unstable. However the damping is often so slight that the motion is very unpleasant and undesirable. It may be described as a yaw and roll to the right, followed by a recovery toward the equilibrium condition, then an overshooting of this condition, and a yaw and roll to the left, then back past the equilibrium attitude, and so on. The name Dutch Roll was attached to the motion because of its similarity to the ice skating figure of the same name. The period is usually of the order of 3 to 15 seconds, so that if the amplitude is appreciable the motion will be very annoying indeed. The damping is increased by large directional stability and small dihedral, and decreased by small directional stability and large dihedral.

---

* Cf. Charles F. Zimmerman, "An Analysis of Lateral Stability in Power-Off Flight with Charts for Use in Design, " *N.A.C.A. Tech. Rep.* 589 (1937).

By comparing (b) and (c) the requirements for spiral stability and high damping of the lateral oscillations are found to be mutually contradictory. The problem of the aerodynamicist in connection with lateral stability is, therefore, to proportion the dihedral and directional stability so that neither requirement is too seriously violated. Most pilots prefer spiral instability to an insufficient damping of the oscillations, so that designers are likely to lean in this direction, i.e., to furnish fairly large directional stability with moderate dihedral, and accept a certain amount of spiral instability.

A few remarks should be made here regarding lateral controls. The rudder is designed to furnish a pure yawing moment and it does this with considerable success. The ailerons should ideally furnish a pure rolling moment, so that the rolling and yawing controls would be entirely independent. Unfortunately most ailerons produce yawing as well as rolling moments. The down-aileron which increases the lift likewise increases the drag while the up-aileron may actually decrease the drag slightly. This means that when the ailerons are deflected so as to give a positive rolling moment (right wing down) they also produce a negative yawing moment (right wing forward). The negative yawing velocity so created gives a negative rolling moment in accordance with the discussion of the previous section ($\partial C_r/\partial r > 0$). This tends to counteract the primary rolling moment caused by the ailerons directly, and may decrease the rolling effect of the ailerons very appreciably. Many methods of reducing or eliminating this "unfavorable" yawing moment have been tried, the most common being a differential rigging of the ailerons so that the up-aileron moves through several times the angle traveled by the down-aileron. An excellent résumé of a very extensive series of investigations conducted by the N.A.C.A. in the field of lateral control devices has been prepared by Weick and Jones.* The reader wishing to pursue the subject further is referred to this work.

5–4. Lateral Motions Above the Stall; Spinning

Probably the most important occurrence connected with the passing of the stall of a normal wing, as the angle of attack increases, is not the decrease in lift but rather the sudden vanishing of the damping in roll, $\partial C_r/\partial p$, which takes place, followed immediately by a strong roll instability. Consider a wing with a rolling velocity as in Fig. 5·4 and suppose that the wing is flying above the stall so that the center of the wing is at an angle of attack like α_1 in Fig. 5·6. This figure represents the curve

---

* Fred E. Weick and Robert T. Jones, "Résumé and Analysis of N.A.C.A. Lateral Control Research," *N.A.C.A. Tech. Rep.* 605 (1937).

of C_L vs. α for any airfoil element along the span of the wing. Now the positive roll shown in Fig. 5·4 makes α for the right wing greater than α_1 while α for the left wing is less than α_1. Near α_1 the lift curve has a negative, instead of a positive slope so that the lift on the right wing is now less, rather than greater, than the lift on the wing. The result is a powerful positive rolling moment which accelerates instead of retarding the small initial roll. It is obvious that a violent lateral instability results from this condition. It is important to discuss the possibility of controlling the inception of the motion resulting from this instability. If the wing is so designed that the central portion stalls first while the tips remain unstalled, then the damping exerted by the tips may, because of their greater lever arms, be sufficient to counteract the negative damping in roll of the stalled central portion. Furthermore in such a case the ailerons will still retain their effectiveness so that the roll may be counteracted with them. The considerable loss of lift over the central portion will then cause the nose of the plane to drop without any roll developing, the angle of attack will decrease as the flying speed picks up, and the recovery from the stall to normal flying conditions can be effected with only a small loss of altitude. On the

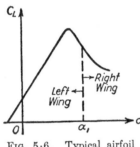

FIG. 5·6. Typical airfoil section lift curve.

other hand if the tips stall before the center the damping in roll disappears and the ailerons become ineffective although very little total lift may have been lost. A violent roll will then occur before the nose has dropped or any other indication of the stall has appeared. At times this effect is so powerful that in less than a second a plane flying steadily and normally whips completely over on its back. If this should occur near the ground recovery would be quite impossible and a serious crash would result. The reason for the current extreme emphasis on good stall characteristics is quite obvious.

The motion which follows such a lateral instability at the stall is usually that known as the spin or tail-spin. The airplane descends rapidly along a helical path having the shape of a corkscrew, for a right-hand spin. During the descent in such a spin the plane is continuously rolling and yawing to the right. The angle of attack at the center of the wing may lie anywhere between about 30° and over 60° so that the plane is definitely stalled. In spite of this the nose is usually pointed towards the earth so steeply that the pilot's instinct is to raise it by pulling back on the stick. This of course has no effect whatever on stopping the spin, since the angle of attack is already far above that for the stall. Sometimes the horizontal tail surfaces still retain enough effectiveness in spite of being stalled so that, if the stick is pushed forward, the angle of attack of the wings can be reduced sufficiently to unstall the airplane which then goes into a dive.   Often, however, this is not possible, the elevators having lost all their effectiveness.   The ailerons also are almost always useless in a spin, so that the rudder is frequently the only control which remains useful.   If the tail surfaces are properly designed, so that the vertical surfaces are not blanketed by the horizontal ones, the rudder retains its effectiveness and can be used to stop the yaw, after which the roll stops itself and the plane goes into a normal dive.   Many airplanes have, however, been built in which the vertical surfaces were completely shielded by the stabilizer and elevator at the large angles of attack corresponding to the spinning motion.   With such airplanes it has often been difficult or impossible to stop a fully developed spin.   The United States government agencies dealing with aircraft now have very stringent spinning tests which must be passed by all small airplanes which might conceivably be spun during normal operation.

Problems

Section 1-1. Pages 1-11.

1. Calculate the total skin friction on a thin flat plate of length L set parallel to the flow direction of a uniform stream of air with velocity U. Assume that the velocity profile has everywhere (over both sides of the plate) the following shape: u increases linearly from the value O at the surface to the free-stream value U at a distance Y from the surface. The plate is 10 ft. wide (perpendicular to L or U). The dimensions are $L = 10$ ft., $U = 100$ ft./sec., $Y = \dfrac{1}{1000}$ ft., and the viscosity coefficient has the value $\mu = 3.7 \times 10^{-7} \dfrac{\text{lb. sec.}}{\text{ft.}^2}$. Give the units in terms of which your answer is given and verify their correctness by dimensional analysis.

2. Calculate C_L, C_D, C_M, and R for a wing of rectangular plan form under the following conditions: wing chord $= 5$ ft., wing span $= 30$ ft., flight velocity $= 300$ ft./sec., lift $= 3000$ lb., drag $= 150$ lb., pitching moment $= 4000$ ft.-lb. Use the standard values of ρ and ν given on p. 17 of the text.

3. Calculate the wing Reynolds numbers of the following airplanes under the indicted conditions:

    (a) Small commercial plane, $c = 4$ ft., at landing and maximum speeds, $V = 35$ and 80 m.p.h.

    (b) Pursuit airplane, $c = 6$ ft., at landing and maximum speeds, $V = 80$ and 400 m.p.h.

    (c) Heavy bomber, $c = 15$ ft., at landing and maximum speeds, $V = 90$ and 350 m.p.h.

Use the standard value of ν as in Prob. 2, and note that in determining R, consistent units must be used.

Section 1-2. Pages 11-17.

4. Determine the lift and drag in pounds and the pitching moment in foot-pounds for an airplane wing having an area of 800 ft.^{2} and a chord of 10 ft., at speeds of 50, 100, 200, and 300 m.p.h., under sea-level standard conditions. The aerodynamic coefficients are given by: $C_L = 0.3$ (independent of R), $C_M = 0.02$ (independent of R), $C_D = 0.01 - R/10^{10}$.

5. Repeat Prob. 4 for a scale model of the airplane tested under the following conditions:

    (a) In the Eiffel tunnel, model scale $= \frac{1}{20}$, wind speed $= 30$ m.p.h., atmospheric pressure.

    (b) In the GALCIT tunnel, model scale $= \frac{1}{10}$, wind speed $= 200$ m.p.h., atmospheric pressure.

(c) In the N.A.C.A. full-scale tunnel, model scale = ½, wind speed = 80 m.p.h., atmospheric pressure.

(d) In the N.A.C.A. variable density tunnel, model scale = ¹⁄₂₅, wind speed = 90 m.p.h., pressure = 20 atmospheres.

6. An airplane is tested in free flight under standard atmosphere conditions. A geometrically similar model is tested at the same attitude in the variable density tunnel pumped up to 20 atmospheres pressure. The wind velocity in the tunnel is ½ the free flight velocity and the two Reynolds numbers are identical. Find the model scale and the ratio of model to full-scale force.

7. Using the data of Fig. 1·12 determine the lift, drag, and pitching moment of the corresponding full-scale airplane with wing area = 551 ft.² and mean wing chord = 8.4 ft., flying at 158 m.p.h. at sea-level at an angle of attack of 4°. Assume all aerodynamic coefficients independent of Reynolds number. Because of the small scale of the figure the accuracy of the results can hardly be greater than two significant figures. (Since the polar of Fig. 1·12 is used in many of the subsequent problems the instructor will find it useful to furnish a large-scale replot of the figure employing a finer coordinate mesh.)

Section 1–3. Pages 17–24.

8. Consider a portion of a wind tunnel as shown in the sketch, the velocity being assumed constant across the entrance section (area S_1, pressure p_1) and

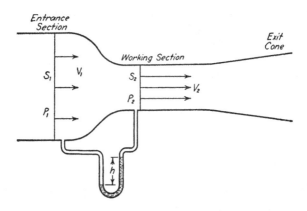

also across the working section (area S_2, pressure p_2). The pressure difference $p_1 - p_2 = \Delta p$ is measured by the difference in level, h, between the two arms of a liquid manometer. If γ is the specific weight of the manometer liquid, then $\Delta p = \gamma h$. Using the continuity equation and the Bernoulli equation (assuming the Bernoulli constant to have the same value throughout the fluid) find an expression for the working section velocity, V_2, in terms of h, the manometer "head"; S_1/S_2, the wind tunnel contraction ratio; γ, the specific weight of the manometer fluid; ρ, the air density. Find the value of V_2 in m.p.h. for the following case: $h = 20$ in., $S_1/S_2 = 4$, $\gamma = 64$ lb./ft.^{3}, $\rho = \rho_0$.

Section 1–4. Pages 24–27.

9. Calculate the dynamic pressure in lb./ft.^{2}, under sea-level conditions, corresponding to velocities of 1, 10, 50, 100, 200, 300, and 400 m.p.h. Neglect compressibility.

Section 1–5. Pages 28–39.

10. Consider the flows of Fig. 1·29, the velocity of the rectilinear flow far from the cylinder being V. For the rectilinear flow of (1) in the figure, the velocity at top and bottom of the cylinder is $2V$. Determine the pressure at top and bottom of the cylinder in the case of combined flows in (3), when the velocity at the cylinder in the pure circulatory flow is 0, $0.1V$, $0.3V$, $0.7V$, and $1.0V$. Refer all pressures to the undisturbed pressure far from the cylinder, and express them in terms of the dynamic pressure $\frac{1}{2}\rho V^2$.

11. A wing moves at a speed of 150 m.p.h. through air of standard density. If the lift per foot of span is 200 lb. determine the circulation around the wing.

12. Determine C_L at an angle of attack of 8° for a symmetrical, thin airfoil section, giving zero lift at 0° angle of attack, taking the airfoil efficiency factor η as 0.90.

Section 1–6. Pages 39–54.

13. Consider two comparable airplanes in level flight under the same conditions: $\rho = \rho_0$, $V = 100$ m.p.h. The two airplanes are identical except for the differences indicated below: $W_1 = W_2 = 10,000$ lb., $S_1 = S_2 = 400$ ft.^{2}, $b_1 = 40$ ft., $b_2 = 60$ ft. Determine the differences in: (a) geometrical angle of attack in degrees, (b) drag in pounds, assuming elliptic lift distribution for both wings.

14. Consider the two airplanes of Prob. 13, assuming that the airfoil sections are symmetrical so that the angle of attack for zero lift is 0°. On the basis of elliptic lift distribution calculate the geometrical angles of attack for the two planes under the conditions of Prob. 13 and compare the difference with that found in the previous problem. Take the airfoil efficiency factor, η, as 0.90.

15. Consider a long-range airplane with the following characteristics: weight = 60,000 lb., wing area = 1250 ft.^{2}, span = 115 ft. Calculate the induced drag for this airplane in level flight at sea level at velocities of 100 m.p.h., 200 m.p.h., 300 m.p.h., and 400 m.p.h.

16. Consider two model wings each having the following characteristics: $S = 6$ ft.^{2}, $\alpha_{L_0} = -1.5°$, $\eta = 0.88$, lift = 75 lb., $V = 125$ m.p.h., elliptic lift distribution. Wing 1 has $\mathcal{R} = 5$, wing 2 has $\mathcal{R} = 8$. Determine α (degrees) and D_i for the two wings and compare.

17. Consider an airplane having an untwisted wing with linear taper in level flight. Given that $V = 200$ m.p.h., $\alpha_{L_0} = -2°$, $\alpha = 4°$, $\rho = \rho_0$, $b = 36$ ft., $\mathcal{R} = 6$, root chord = 7 ft., $C_{D_0} = 0.01$, calculate the total wing drag assuming non-elliptical lift distribution.

Section 1–7. Pages 54–65.

18. For a flat plate set parallel to a given airstream the boundary layer thickness 10 ft. downstream from the leading edge is 0.15 in. Determine the boundary layer thickness 1 ft., 3 ft., and 7 ft. from the leading edge.

Section 2–1. Pages 66–73.

19. Compare α_{L_0}, a_0, $C_{L_{max}}$, $C_{D_{0_{min}}}$, and C_{M_0} for the following airfoils: 0021, 2412, 2418, 4412, 23012. Use Table 3, *N.A.C.A. Tech. Rep.* 669.

Section 2–3. Pages 77–84.

Note: In solving the biplane problems below use must be made of one of the existing sets of charts for the Munk span factor k.

20. Find the equivalent monoplane spans for the following biplanes:

(a) $b_1 = 45'$, $b_2 = 32'$, $G = 6'9''$, $c_1 = 7'$, $c_2 = 5'6''$
(b) $b_1 = 30'$, $b_2 = 25'$, $G = 4'6''$, $c_1 = 4'$, $c_2 = 4'$
(c) $b_1 = 80'$, $b_2 = 64'$, $G = 10'$, $c_1 = 10'$, $c_2 = 8'$

21. Find the induced drag of the following biplane flying at 150 m.p.h. under sea-level standard conditions: $W = 7000$ lb., $b_1 = 42'$, $b_2 = 30'$, $G = 5'$, $c_1 = 6'6''$, $c_2 = 5'$.

Section 2–4. Pages 84–89.

22. An airplane of 4000 lb. gross weight has a wing area of 150 sq. ft. Using the data given on pp. 86–87 of the text, find the sea-level landing speed (stalling speed) for the following cases:

(a) No high-lift device, N.A.C.A. 23012 airfoil.
(b) No high-lift device, Clark Y airfoil.
(c) Plain flap on 23012.
(d) Handley-Page slot.
(e) Fowler flap.
(f) Handley-Page slot and Fowler flap.

Section 2–5. Pages 89–98.

23. An airplane has 1000 ft.^{2} wing area. The wing profile drag coefficient is $C_{D_0} = 0.0095$. The projected frontal area of the fuselage is 90 ft.^{2} and its proper drag coefficient is $C_{D_\pi} = 0.085$. The tail surface area is 200 ft.^{2} and its proper drag coefficient is 0.0072. The nacelles add an equivalent parasite area of $f = 3.8$ ft.^{2} Find the parasite drag coefficient of the complete airplane.

24. Given standard conditions, a thin airfoil with chord = 6 ft., and assuming that the wetted area = $2S$, find the skin friction per foot of span at velocities of 10, 50, 150, and 400 m.p.h. for the following cases:

(a) Laminar boundary layer flow.

(b) Turbulent boundary layer flow.

(c) Boundary layer flow in which transition occurs within our range of Reynolds numbers. Use the middle transition curve of Fig. 2·31.

Section 3–1. Pages 99–103.

25. A pilot flying in an airplane weighing 10,000 lb. and having 300 ft.² wing area and 7 ft. chord obtains the following readings on his instruments: h_p = 20,350 ft., T = 31.6° F. (outside air temperature), V_i = 185 m.p.h. Find the corresponding values of the following quantities: p(mm. Hg), h_d, σ, $\sqrt{\sigma}$, V_{true} C_L, R.

26. Determine the propulsive efficiency of a propeller to which an engine is delivering 1000 hp. and which is furnishing 1500 lb. thrust at a forward speed of 200 m.p.h.

Section 3–2. Pages 103–106.

27. Calculate the maximum rate of climb (ft./min.) for a 15,000-lb. airplane whose maximum excess horsepower is 600.

28. Determine the minimum time to climb from sea level to 20,000 ft. for an airplane whose maximum rate of climb is given by C (ft./min.) = 1500 − ⅟₂₀h (ft.).

Section 3–3. Pages 106–114.

Note: Many of the performance problems of this and the following sections are based on data (airplane polar, engine chart, and propeller charts) given in the text. These problems may, of course, be varied by furnishing other analogous data.

29. Plot a curve of Pr_i(hp.) vs. V_i (m.p.h.) for the airplane whose polar is given in Fig. 3.8, assuming C_D independent of R and taking W = 17,500 lb., S = 551 ft.², b = 65.5 ft. Cover the range from V_i = 0 to 300 m.p.h.

30. Given an airplane with an aspect ratio of 6, and a polar such that C_L = 0.2, C_D = 0.0380; C_L = 0.8, C_D = 0.0730, find $Æ_e$, e, $C_{D_{p_e}}$, $(L/D)_{\max}$.

31. From the curve of Prob. 29 find the minimum power required at sea level, 10,000 ft., 20,000 ft., 30,000 ft. in the standard atmosphere.

32. Repeat Prob. 30 for the airplane of Prob. 29, fitting the polar with a parabola at C_L = 0.2 and 0.8.

33. Determine the span and parasite loadings (engineering) of the airplane of Prob. 29, using the results of Prob. 32.

34. Using equation 3·28 and the loadings of Prob. 33 plot the "ideal" indicated power-required curve for the airplane of Prob. 29. Compare the result with the corresponding curve of Prob. 29.

35. Transform equation 3·15 so as to give $(L/D)_{\max}$ in terms of the span and parasite loadings. Find the expression giving indicated speed for $(L/D)_{\max}$ in terms of these loadings.

Section 3–4. Pages 114–119.

36. Determine the manifold pressure required for the G 102 engine to produce 600 b.hp. at 1900 r.p.m., with carburetor air temperature = standard, $h_p = 5000$ ft., zero ram. Use the engine chart of Fig. 3·10.

37. Consider the airplane of Prob. 29 with two Wright G 102 engines (corresponding to Fig. 3·10) each driving an 11-ft., three-blade, constant-speed propeller (corresponding to Figs. 3·12 and 3·13). Determine J, C_P, C_T, η, β under the following operating conditions (standard atmosphere):

(a) $h = 7000$ ft., $N_{engine} = 2300$, $MP = 35.4$ in. Hg, $V = 265$ m.p.h.
(b) $h = 15,000$ ft., $N_{engine} = 2000$, full throttle, $V = 240$ m.p.h.

38. For the airplane of Prob. 37, with the engine at 2300 r.p.m. delivering 900 hp. below critical altitude and operating at full throttle above critical altitude, determine power available as a function of velocity for sea level and altitudes of 7000 ft., 15,000 ft., 22,000 ft.

Section 3–5. Pages 119–130.

39. Plot the results of Prob. 38 on the same indicated power-indicated velocity diagram as was used in Prob. 29, and determine graphically the elements of the classical performance problem for the chosen altitudes.

40. Plot the results of Prob. 39 on a classical performance diagram like Fig. 3·6.

41. Using the Composite Performance Chart determine V_0, V_{mp}, and C_{mp} for the airplane of Prob. 38 at critical altitude assuming $R_p = 1$.

42. For the above airplane, but with the actual variation of R_p with V, find V_c and C at critical altitude using Rockefeller's method. Repeat the calculations for values of R_R 10 per cent greater and 10 per cent less than the value given by the empirical curve of the Composite Performance Chart, and compare the three rates of climb.

43. Calculate by Rockefeller's method the same elements of the performance problem as were determined graphically in Prob. 39 and compare the two sets of results.

44. For the above airplane find the effect on V_{max} and C at critical altitude, and the effect on absolute ceiling, of the following modifications considered separately: 5 per cent increase in W, 5 per cent increase in f, 5 per cent decrease in P, 5 per cent decrease in b.

Section 3–6. Pages 130–136.

45. Given a large airplane with the following characteristics, and assuming that the airplane flies at a $C_L = 0.42$, determine the range at sea level by use of the Breguet formula: loaded gross weight = 50,000 lb., fuel capacity = 1900 gallons (fuel weight, 6 lb. per gallon), $S = 1160$ ft.^{2}, $b = 98$ ft., $e = 0.93$, $C_{DP_i} = 0.027$, $\eta = 0.85$, $c = 0.50$ lb./b.hp./hr.

46. Determine the take-off distance for a 40,000-lb. airplane assuming that the effective thrust is constant and that the airplane leaves the ground at a velocity 15 per cent greater than the stalling velocity. Take $C_{L\max}$ from the polar of Fig. 1·44, $S = 950$ ft.^{2}, $T_e = 11,500$ lb.

Section 4–2. Pages 141–143.

47. Using standard N.A.C.A. airfoil characteristic data (e.g., Table III, N.A.C.A. Tech. Rep. 669) plot curves showing C_{Mw} as function of C_L for airfoil sections 2412, 4415, 23012 with $\delta/c = 0.2, 0.3, 0.4$. Determine the tail moment coefficient required for trim at $C_L = 0.8$ for all the above cases. Take $C_{M0} =$ constant and the c.g. on the chord line.

Section 4–3. Pages 143–147.

48. Given the following airplane characteristics: wing—23012 section, incidence to thrust line = $2°$, $\mathcal{R} = 7$, $S = 500$ ft.^{2}, $c = 8.45$ ft.; tail—0009 section, $S_t = 75$ ft.^{2}, $\mathcal{R}_t = 3.8$, $l = 24$ ft., $\eta_t = 0.70$, $a_0 = 5.7$ for wing and tail, $C_{M0} =$ constant, elevator neutral. Present equations and curves giving C_{Mt} vs. C_L for stabilizer settings relative to thrust axis of $+2°, -2°, -6°$.

49. Repeat Prob. 48 assuming that all dimensionless ratios and other quantities remain as above except for the wing and tail aspect ratios. Consider the following two cases:

(a) $\mathcal{R}_w = 3.5$, $\mathcal{R}_t = 3.8$.
(b) $\mathcal{R}_w = 7.0$, $\mathcal{R}_t = 1.9$.

Section 4–4. Pages 148–151.

50. Consider the skeleton airplane of Prob. 48 with c.g. on the chord line 28 per cent c aft of the leading edge. Determine the stabilizer setting limits required to permit trim throughout the flying range from $C_L = 0.2$ to $C_L = 1.4$ and plot curves of C_M vs. C_L for these stabilizer settings.

51. Repeat Prob. 50 but take the c.g. 3 ft. below the chord line instead of on it and consider the airplane as having a fuselage and wing engine nacelles. Neglect the curvature effect of the vertical displacement of c.g. and use the average empirical values given in the text for fuselage and nacelle effects on stability.

52. Consider the complete airplane of Prob. 51 and assume the tail surface to be rectangular with an elevator chord of 1.75 ft. Determine the stabilizer setting for trim at $C_L = 0.5$ with elevator neutral. Plot curves of C_M vs. C_L for this stabilizer setting with elevator at $+15°, 0°, -20°$.

INDEX

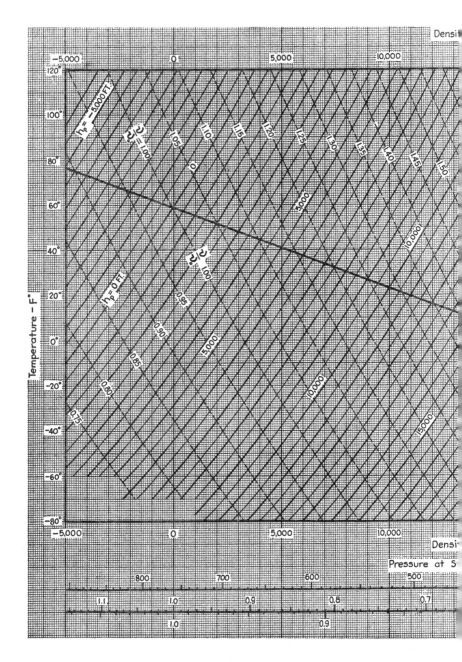

Density

Temperature – F°

$h_p = -5000$ FT

$\frac{\rho}{\rho_0} = 1.00$

1.05
1.10
1.15
1.20
1.25
1.30
1.35
1.40
1.45
1.50

5000
10,000

$\frac{\rho}{\rho_0} = 1.00$

0.95
0.90
0.85
0.80
0.75

$h_p = 0$ FT

5000
10,000
15,000

Density

Pressure at S

800   700   600   500

1.1   1.0   0.9   0.8   0.7

1.0   0.9

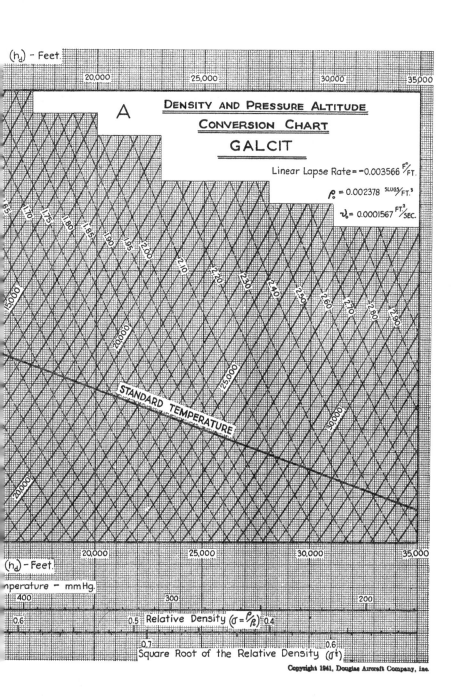

(h_d) – Feet.

20,000   25,000   30,000   35,000

A

DENSITY AND PRESSURE ALTITUDE
CONVERSION CHART
GALCIT

Linear Lapse Rate = −0.003566 $\frac{F°}{FT.}$

ρ_0 = 0.002378 $^{SLUGS}/_{FT.^3}$

ν_0 = 0.0001567 $^{FT.^2}/_{SEC.}$

1.65 1.70 1.75 1.80 1.85 1.90 1.95 2.00 2.10 2.20 2.30 2.40 2.50 2.60 2.70 2.80 2.90

15,000

20,000

25,000

STANDARD TEMPERATURE

30,000

20,000

(h_d) – Feet.

nperature − mmHg.
400   300   200

0.6   0.5   Relative Density ($\sigma = \rho/\rho_0$)   0.4

0.7   Square Root of the Relative Density ($\sigma^{\frac{1}{2}}$)   0.6

Copyright 1941, Douglas Aircraft Company, Inc.

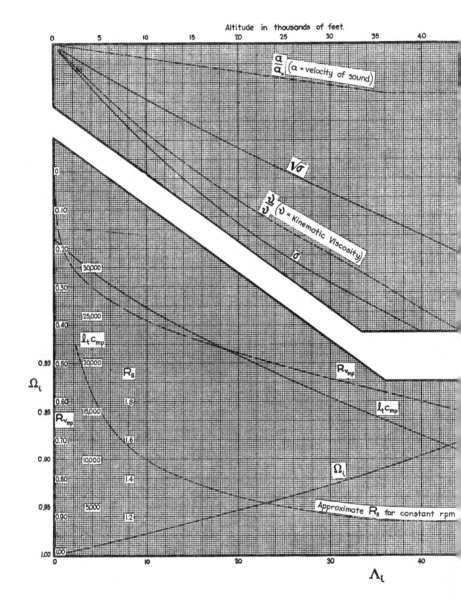

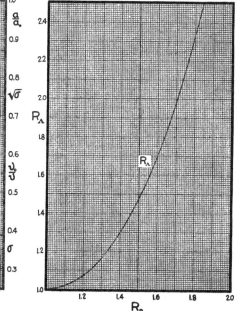

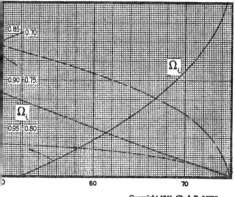

B

COMPOSITE PERFORMANCE CHART

GALCIT

PROCEDURE FOR PERFORMANCE CALCULATION

Given W, b_e, f, $N_o d$, h, $P_o(h)$

Compute $l_s = \dfrac{W}{b_e^2}$, $l_p = \dfrac{W}{f}$, $\eta_o l t_i = \dfrac{W}{\sqrt{\sigma_o} P_o}$ (σ_o corresponds to h)

Maximum Level Speed at Altitude h (σ_o)

1) Assume η_o, compute l_{t_i} and $\Lambda_i = \dfrac{l_s l_{t_i}}{(l_p / l_{t_i})^{\frac{1}{3}}}$

2) Find $\Omega_i (\Lambda_i)$ from comp. perf. chart

3) Compute $V_o = \dfrac{V_{i_o}}{\sqrt{\sigma_o}} = 52.73 \dfrac{(l_p / l_{t_i})^{\frac{1}{3}}}{\sqrt{\sigma_o}} \Omega_i$ (m.p.h.)

4) Compute $J_o = \dfrac{88 V_o}{N_o d}$, $C_{P_o} = \dfrac{0.5 (P_o / 1000)}{\sigma_o N_o^3 d^5} \times 10^{14}$

5) Find $\eta_o (J_o, C_{P_o})$ from propeller chart

6) If η_o differs from that assumed in 1) repeat until agreement is reached.

Maximum Rate of Climb at Altitude h

1) Find $R_R (\Lambda_i)$ and $R_{V_{mp}} (\Lambda_i)$ from comp. perf. chart

2) Compute $R_{V_c} = R_R R_{V_{mp}}$

3) Compute $J_c = R_{V_c} J_o$ and find $\eta_c (J_c, C_{P_o})$ from propeller chart

4) Find $R_\Lambda (R_R)$ from comp. perf. chart

5) Compute $l_{t_c} = \sqrt{\sigma_o} \dfrac{\eta_o}{\eta_c} l_{t_i}$ and $\Lambda_{i_c} = \left(\dfrac{\eta_o}{\eta_c}\right)^{\frac{4}{3}} \Lambda_i R_\Lambda$

6) Find $l_t C_{mp} (\Lambda_{i_c})$ from comp. perf. chart

7) Compute $C(h) = \dfrac{l_t C_{mp}}{l_{t_c}}$ (ft. per min.)

8) Compute $V_c = R_{V_c} V_o$

9) To verify the correctness of the R_R assumed the procedure may be repeated for other values of R_R slightly greater and smaller than that chosen in 1).

Copyright 1941, Clark B. Millikan

A CATALOG OF SELECTED
DOVER BOOKS
IN SCIENCE AND MATHEMATICS

Engineering

FUNDAMENTALS OF ASTRODYNAMICS, Roger R. Bate, Donald D. Mueller, and Jerry E. White. Teaching text developed by U.S. Air Force Academy develops the basic two-body and n-body equations of motion; orbit determination; classical orbital elements, coordinate transformations; differential correction; more. 1971 edition. 455pp. 5 3/8 x 8 1/2. 0-486-60061-0

INTRODUCTION TO CONTINUUM MECHANICS FOR ENGINEERS: Revised Edition, Ray M. Bowen. This self-contained text introduces classical continuum models within a modern framework. Its numerous exercises illustrate the governing principles, linearizations, and other approximations that constitute classical continuum models. 2007 edition. 320pp. 6 1/8 x 9 1/4. 0-486-47460-7

ENGINEERING MECHANICS FOR STRUCTURES, Louis L. Bucciarelli. This text explores the mechanics of solids and statics as well as the strength of materials and elasticity theory. Its many design exercises encourage creative initiative and systems thinking. 2009 edition. 320pp. 6 1/8 x 9 1/4. 0-486-46855-0

FEEDBACK CONTROL THEORY, John C. Doyle, Bruce A. Francis and Allen R. Tannenbaum. This excellent introduction to feedback control system design offers a theoretical approach that captures the essential issues and can be applied to a wide range of practical problems. 1992 edition. 224pp. 6 1/2 x 9 1/4. 0-486-46933-6

THE FORCES OF MATTER, Michael Faraday. These lectures by a famous inventor offer an easy-to-understand introduction to the interactions of the universe's physical forces. Six essays explore gravitation, cohesion, chemical affinity, heat, magnetism, and electricity. 1993 edition. 96pp. 5 3/8 x 8 1/2. 0-486-47482-8

DYNAMICS, Lawrence E. Goodman and William H. Warner. Beginning engineering text introduces calculus of vectors, particle motion, dynamics of particle systems and plane rigid bodies, technical applications in plane motions, and more. Exercises and answers in every chapter. 619pp. 5 3/8 x 8 1/2. 0-486-42006-X

ADAPTIVE FILTERING PREDICTION AND CONTROL, Graham C. Goodwin and Kwai Sang Sin. This unified survey focuses on linear discrete-time systems and explores natural extensions to nonlinear systems. It emphasizes discrete-time systems, summarizing theoretical and practical aspects of a large class of adaptive algorithms. 1984 edition. 560pp. 6 1/2 x 9 1/4. 0-486-46932-8

INDUCTANCE CALCULATIONS, Frederick W. Grover. This authoritative reference enables the design of virtually every type of inductor. It features a single simple formula for each type of inductor, together with tables containing essential numerical factors. 1946 edition. 304pp. 5 3/8 x 8 1/2. 0-486-47440-2

THERMODYNAMICS: Foundations and Applications, Elias P. Gyftopoulos and Gian Paolo Beretta. Designed by two MIT professors, this authoritative text discusses basic concepts and applications in detail, emphasizing generality, definitions, and logical consistency. More than 300 solved problems cover realistic energy systems and processes. 800pp. 6 1/8 x 9 1/4. 0-486-43932-1

THE FINITE ELEMENT METHOD: Linear Static and Dynamic Finite Element Analysis, Thomas J. R. Hughes. Text for students without in-depth mathematical training, this text includes a comprehensive presentation and analysis of algorithms of time-dependent phenomena plus beam, plate, and shell theories. Solution guide available upon request. 672pp. 6 1/2 x 9 1/4. 0-486-41181-8

Browse over 9,000 books at www.doverpublications.com

HELICOPTER THEORY, Wayne Johnson. Monumental engineering text covers vertical flight, forward flight, performance, mathematics of rotating systems, rotary wing dynamics and aerodynamics, aeroelasticity, stability and control, stall, noise, and more. 189 illustrations. 1980 edition. 1089pp. 5 5/8 x 8 1/4.   0-486-68230-7

MATHEMATICAL HANDBOOK FOR SCIENTISTS AND ENGINEERS: Definitions, Theorems, and Formulas for Reference and Review, Granino A. Korn and Theresa M. Korn. Convenient access to information from every area of mathematics: Fourier transforms, Z transforms, linear and nonlinear programming, calculus of variations, random-process theory, special functions, combinatorial analysis, game theory, much more. 1152pp. 5 3/8 x 8 1/2.   0-486-41147-8

A HEAT TRANSFER TEXTBOOK: Fourth Edition, John H. Lienhard V and John H. Lienhard IV. This introduction to heat and mass transfer for engineering students features worked examples and end-of-chapter exercises. Worked examples and end-of-chapter exercises appear throughout the book, along with well-drawn, illuminating figures. 768pp. 7 x 9 1/4.   0-486-47931-5

BASIC ELECTRICITY, U.S. Bureau of Naval Personnel. Originally a training course; best nontechnical coverage. Topics include batteries, circuits, conductors, AC and DC, inductance and capacitance, generators, motors, transformers, amplifiers, etc. Many questions with answers. 349 illustrations. 1969 edition. 448pp. 6 1/2 x 9 1/4.

0-486-20973-3

BASIC ELECTRONICS, U.S. Bureau of Naval Personnel. Clear, well-illustrated introduction to electronic equipment covers numerous essential topics: electron tubes, semiconductors, electronic power supplies, tuned circuits, amplifiers, receivers, ranging and navigation systems, computers, antennas, more. 560 illustrations. 567pp. 6 1/2 x 9 1/4.   0-486-21076-6

BASIC WING AND AIRFOIL THEORY, Alan Pope. This self-contained treatment by a pioneer in the study of wind effects covers flow functions, airfoil construction and pressure distribution, finite and monoplane wings, and many other subjects. 1951 edition. 320pp. 5 3/8 x 8 1/2.   0-486-47188-8

SYNTHETIC FUELS, Ronald F. Probstein and R. Edwin Hicks. This unified presentation examines the methods and processes for converting coal, oil, shale, tar sands, and various forms of biomass into liquid, gaseous, and clean solid fuels. 1982 edition. 512pp. 6 1/8 x 9 1/4.   0-486-44977-7

THEORY OF ELASTIC STABILITY, Stephen P. Timoshenko and James M. Gere. Written by world-renowned authorities on mechanics, this classic ranges from theoretical explanations of 2- and 3-D stress and strain to practical applications such as torsion, bending, and thermal stress. 1961 edition. 560pp. 5 3/8 x 8 1/2.   0-486-47207-8

PRINCIPLES OF DIGITAL COMMUNICATION AND CODING, Andrew J. Viterbi and Jim K. Omura. This classic by two digital communications experts is geared toward students of communications theory and to designers of channels, links, terminals, modems, or networks used to transmit and receive digital messages. 1979 edition. 576pp. 6 1/8 x 9 1/4.   0-486-46901-8

LINEAR SYSTEM THEORY: The State Space Approach, Lotfi A. Zadeh and Charles A. Desoer. Written by two pioneers in the field, this exploration of the state space approach focuses on problems of stability and control, plus connections between this approach and classical techniques. 1963 edition. 656pp. 6 1/8 x 9 1/4.

0-486-46663-9

Browse over 9,000 books at www.doverpublications.com

Physics

THEORETICAL NUCLEAR PHYSICS, John M. Blatt and Victor F. Weisskopf. An uncommonly clear and cogent investigation and correlation of key aspects of theoretical nuclear physics by leading experts: the nucleus, nuclear forces, nuclear spectroscopy, two-, three- and four-body problems, nuclear reactions, beta-decay and nuclear shell structure. 896pp. 5 3/8 x 8 1/2. 0-486-66827-4

QUANTUM THEORY, David Bohm. This advanced undergraduate-level text presents the quantum theory in terms of qualitative and imaginative concepts, followed by specific applications worked out in mathematical detail. 655pp. 5 3/8 x 8 1/2. 0-486-65969-0

ATOMIC PHYSICS AND HUMAN KNOWLEDGE, Niels Bohr. Articles and speeches by the Nobel Prize–winning physicist, dating from 1934 to 1958, offer philosophical explorations of the relevance of atomic physics to many areas of human endeavor. 1961 edition. 112pp. 5 3/8 x 8 1/2. 0-486-47928-5

COSMOLOGY, Hermann Bondi. A co-developer of the steady-state theory explores his conception of the expanding universe. This historic book was among the first to present cosmology as a separate branch of physics. 1961 edition. 192pp. 5 3/8 x 8 1/2. 0-486-47483-6

LECTURES ON QUANTUM MECHANICS, Paul A. M. Dirac. Four concise, brilliant lectures on mathematical methods in quantum mechanics from Nobel Prize-winning quantum pioneer build on idea of visualizing quantum theory through the use of classical mechanics. 96pp. 5 3/8 x 8 1/2. 0-486-41713-1

THE PRINCIPLE OF RELATIVITY, Albert Einstein and Frances A. Davis. Eleven papers that forged the general and special theories of relativity include seven papers by Einstein, two by Lorentz, and one each by Minkowski and Weyl. 1923 edition. 240pp. 5 3/8 x 8 1/2. 0-486-60081-5

PHYSICS OF WAVES, William C. Elmore and Mark A. Heald. Ideal as a classroom text or for individual study, this unique one-volume overview of classical wave theory covers wave phenomena of acoustics, optics, electromagnetic radiations, and more. 477pp. 5 3/8 x 8 1/2. 0-486-64926-1

THERMODYNAMICS, Enrico Fermi. In this classic of modern science, the Nobel Laureate presents a clear treatment of systems, the First and Second Laws of Thermodynamics, entropy, thermodynamic potentials, and much more. Calculus required. 160pp. 5 3/8 x 8 1/2. 0-486-60361-X

QUANTUM THEORY OF MANY-PARTICLE SYSTEMS, Alexander L. Fetter and John Dirk Walecka. Self-contained treatment of nonrelativistic many-particle systems discusses both formalism and applications in terms of ground-state (zero-temperature) formalism, finite-temperature formalism, canonical transformations, and applications to physical systems. 1971 edition. 640pp. 5 3/8 x 8 1/2. 0-486-42827-3

QUANTUM MECHANICS AND PATH INTEGRALS: Emended Edition, Richard P. Feynman and Albert R. Hibbs. Emended by Daniel F. Styer. The Nobel Prize–winning physicist presents unique insights into his theory and its applications. Feynman starts with fundamentals and advances to the perturbation method, quantum electrodynamics, and statistical mechanics. 1965 edition, emended in 2005. 384pp. 6 1/8 x 9 1/4. 0-486-47722-3

Browse over 9,000 books at www.doverpublications.com